THE EVOLUTION OF CLOCKWORK

THE EVOLUTION OF CLOCKWORK

With a special section on
The Clocks of Japan

Together with a Comprehensive
Bibliography of Horology

By

J. DRUMMOND ROBERTSON

101 ILLUSTRATIONS

Republished S.R. Publishers Ltd., 1972
First published 1931

Reproduced 1972 S.R. Publishers Limited,
East Ardsley, Wakefield,
Yorkshire, England,
by kind permission of the
copyright holder.

ISBN 0 85409 748 1

Please address all enquiries to S.R. Publishers Ltd.
(address as above)

Reprinted by Scolar Press Ltd.,
Menston, Yorkshire, U.K.

THE EVOLUTION OF CLOCKWORK

Fifteenth-century Clock from an Italian Tarsia Panel
By courtesy of the Victoria and Albert Museum

THE EVOLUTION OF CLOCKWORK

WITH A SPECIAL SECTION ON

THE CLOCKS OF JAPAN,

FULLY ILLUSTRATED FROM THE AUTHOR'S COLLECTION

TOGETHER WITH

A COMPREHENSIVE BIBLIOGRAPHY OF HOROLOGY

COVERING OVER SIX HUNDRED AUTHORS

BY

J. DRUMMOND ROBERTSON

101 ILLUSTRATIONS

Luna rotat velox, currit Sol, Hora volutat;
Heu! velis aut nolis: sic tua vita volat.

CASSELL & COMPANY, LTD.
LONDON, TORONTO, MELBOURNE & SYDNEY

First published 1931

Printed in Great Britain

PREFACE

A GLANCE at the bibliography at the end of this volume may well suggest the question whether there is room for another work on horology. My answer must be that the present book is not concerned with the practical side of clockmaking, but is an endeavour to illustrate some of the historical aspects of the subject. The following chapters might therefore be described as "Horological Essays."

It may safely be said that the origin of the clock as an instrument of time measurement is lost in the mists of the so-called "dark ages," and will never with certainty be known. Its predecessors, the sundial and the clepsydra, go back to remote antiquity ; but with them I am not here concerned. Such records as have come down to us of the clock, as we generally understand the word, are not sufficiently precise to enable us to attribute the birth of this great invention to any author or place. Its evolution, therefore, proceeds from unknown beginnings.

My object, then, has been to trace in broad outline its development through the "Middle Ages," where we rest on a fairly certain historical basis, down to the adoption of the pendulum and the balance spring as regulating agents in the latter half of the seventeenth century. We then take leave of the traditions of mediæval practice, and at once enter upon the new era of modern horology and exact time measurement. On that subject, of the writing of books there has been no end, to which my bibliography affords eloquent testimony ; and it is not within the scope of this work to add to their number.

A few words must be said in justification of the lengthy chapter on the Pendulum, and of the resuscitation of the controversy as to whom belongs the honour of its first application to clockwork, which has raged for wellnigh three centuries. I think that I may claim that the pros and cons have never yet been presented in such detail, supported by original documents and drawings : in England, at any rate, nothing of the kind has been attempted.

For much of the information concerning Galileo's clock I am greatly indebted to Prof. Pietro Pagnini of Florence, who most kindly made research on my behalf in the Medicean Archives, furnishing lengthy extracts from them, as also a full-sized photograph of the original drawing, which is reproduced on a smaller scale, facing the Appendix to Chapter VI.

I owe to the courtesy of the Bodleian Library the reproduction of the portrait of Galileo by Sustermans, which is of particular interest as having been painted at the very time at Arcetri, when the blind old man was devising with his son the mechanism of his pendulum clock.

Especial thanks are due to Dr. J. A. Vollgraff, the present learned editor of such works and correspondence of Huygens as still remain to complete this monumental undertaking. As the result of a visit to the University of Leiden, to which Huygens bequeathed all his vast accumulation of papers, and a subsequent voluminous correspondence with Dr. Vollgraff, I have been enabled to present a fairly faithful picture of Huygens' ceaseless activities in the development of his pendulum clock, and in his attempts to solve the problem of the longitude at sea, which have never yet been fully related.

Without Dr. Vollgraff's assistance the chapters relating to Huygens could not have been written. A first instalment dealing with Huygens' horological activities in the fullest detail—" Christian Huygens : L'Horloge à pendule de 1656 à 1666 "—has just been published by him, to which I would refer interested readers.

I owe to the Gemeente Museum at The Hague the charming and little-known portrait of Huygens at the age of forty-one by Caspar Netscher.

I have also to thank Mr. Percy Webster for photographs of three of the mediæval clocks in his wonderful collection, as also for valuable hints, suggested by his unrivalled knowledge of everything connected with horology.

The illustration of the Dover Castle clock is given by permission of the Director of the Science Museum.

The section dealing with Japanese clocks is the first attempt to treat this interesting branch of horology comprehensively. Apart from a chapter in M. Planchon's work "L'Horloge," and a few occasional articles, not always accurate, in the technical English and Continental journals, the subject is one which has not received the attention that it deserves ; and I trust that the illustrations, derived entirely from my own collection, will serve to show that we have here horological matter of no small interest.

Whether viewed from the standpoint of primitive time measurement, of the retention of the mediæval bar-balance, of ingenious invention, or of artistic treatment, these clocks offer a fascinating study.

The Bibliography which concludes this book is also the first attempt in England to present a compendious survey of the literature of horology. In 1897 a lengthy list of works up to that date was published in Germany by Herr M. Loeske, which has enabled me to fill many gaps in my catalogue.

Mr. C. A. Ilbert has been kind enough to read the proofs of this section, and to collate the catalogue of the numerous works and pamphlets in the Library of the Institute of Civil Engineers, resulting in several corrections and additions.

The works enumerated in this bibliography have been limited, with few exceptions, to such as deal with the historical and technical aspects of clockwork. In some cases articles which have appeared in periodicals, when of sufficient interest, have been included, as also their reprints to be found in the public libraries. For many other articles of less general interest reference may be made to the List of Works on Horology issued by the Library of the Patent Office in 1912, a catalogue which unfortunately is not arranged alphabetically under its different sections. I have to thank the librarian for a list of works added to the library since the above date.

J. D. R.

COMRIE LODGE,
TORQUAY.
September 1931.

CONTENTS

THE EVOLUTION OF CLOCKWORK

THE CLOCKS OF JAPAN

LIST OF ILLUSTRATIONS

THE CLOCKS OF JAPAN

"LANTERN" CLOCKS

"BRACKET" CLOCKS

"PILLAR" CLOCKS

FANCY CLOCKS

THE EVOLUTION OF CLOCKWORK

INTRODUCTION

MEASUREMENT OF TIME

So accustomed are we, and so accustomed have our forefathers been for many generations, to our present system of dividing the day into twenty-four hours of equal length, with their subdivisions into minutes and seconds, that it is not altogether easy to realize that this method of time measurement is not only one of comparatively modern usage, but that it is based upon a compromise.

Two facts would at once force themselves upon the observation of the most primitive races of mankind—the sun appearing in the east, travelling across the heavens and rising higher and higher in its course, then gradually descending and disappearing in the west. The next fact impressed upon the savage mind would be that the sun, when it reached the highest point in its daily progress, stood always at much about the same point above some outstanding object in the landscape in relation to the tribal settlement. The constant recurrence of these phenomena would give rise to the conception of a day ; the day in turn dividing itself naturally into light and darkness, or the respective periods between sunrise and sunset and sunset and sunrise. It would also become apparent that the position in which the sun rose and set varied with the season of the year ; whereas its position at one moment during daylight remained fairly constant. An upright tree would cast its shadow day by day in the same direction, and thus the third factor in the daily round of midday or noon would obtain recognition.

Such a day, divided into light and darkness, the periods varying in length with the seasons, is known as the " natural day."

The subdivision of these periods would not trouble the primitive mind ; but with the growth of man's intelligence and his constant observation of the heavenly bodies, he would, in the course of numberless centuries, be led to the adoption of some system for designating time and seasons.

The origin of such a system or systems amongst different races before they emerged from barbarism is necessarily impossible of determination. It is not until we leave prehistoric man behind and come down to the advanced civilization of such peoples as the Sumerians, Egyptians and Chinese that we have any means of judging the results of the empiric knowledge accumulated and handed down through the long dark ages of human progress.

Beside the phenomena of the solar day, it is evident that primitive man must have noted the waxing and waning of the moon. If he commenced to make notches in a stick at full moon in the height of summer or depth of winter he would find that, after marking twelve such notches, each corresponding with the return of the full moon, the season was much the same as when he commenced.

Here then we have the basic idea of a year divided into twelve moons or months, which has descended to our time through the countless ages. Nay, may we not suppose that in this measurement of time, the first measurement to which the untutored mind gave thought, originated the duodecimal system of calculation (instead of the obvious decimal method of counting with the ten fingers)—a system which was later on to be applied to the duodecimal subdivision of standards of weight and length, when the early architect and the trader experienced the necessity for creating smaller units for carrying on their work ?

But our primitive man would not only notch his stick at each full moon. He would want to know how often the sun would rise between full moon and full moon ; and from another stick, marked day by day, he would find that thirty notches were required to make the tale complete. How soon he would discover that his tally of thirty notches was inexact, or that his lunar year of twelve moons ceased to agree with the season at which his count of the years began, it is impossible to say. It is not until a people has advanced to so high a state of civilization as to have been able to leave records for posterity of its accumulated knowledge of mathematics and astronomy that we are in a position to judge how these problems had been dealt with, and in what way the discrepancies had been adjusted.

Naturally, we must turn to the two oldest civilizations of historic times, the Egyptian and Babylonian.

For a people whose religion, like that of the ancient Egyptians, was based on worship of the sun, it was but natural to adopt a solar year. They had, however, already inherited the system of reckoning months of thirty days each, and it was therefore necessary to harmonize this year of 360 days with the solar year of 365 days, which it was then supposed to contain. This adjustment was effected more than three thousand years before our era by adding five epagomenal days after the end of the last month. To these five days were attributed the births respectively of Osiris, Horus, Set, Isis and Nephthys. The civil day was divided into twenty-four hours, and again into minutes and seconds ; and, like our own, it commenced at midnight.

The Babylonian Empire was founded on the conquest by a Semitic race of the ancient non-Semitic race, known as the Sumerians, a people to whom a Tartar or Turanian origin has been attributed. They had reached at the time of their conquest a very high level of civilization, and had already developed the system of cuneiform writing, which was adopted by their conquerors, with the rest of their advanced culture. This form of writing was based upon picture-characters, like the primitive Chinese writing, and perhaps both were founded on a common origin before the emigration westwards of the Tartar tribes.

Amongst the tablets in the British Museum there are many which contain astrological forecasts and astronomical observations, some of them written in Sumerian ; but these date from the later period of Babylonian history in the third century B.C.

At this time it would appear that the astronomical year was divided into twelve months containing twenty-nine and thirty days, the lunar year being adjusted by the insertion of intercalary months.[1]

It is believed that the commencement of the day for astronomical purposes was reckoned from midnight, the whole day being divided into six parts, each containing sixty subdivisions. There appears, however, to have remained in popular use a much older system, by which the day commenced at sunset, with twelve divisions, each " hour " being again subdivided into thirty parts.[2]

It is not known whether the Chinese derived their measurement of time from the Sumerians, but it is significant that this primitive system of twelve divisions, beginning at sunset, was that which obtained in China, and was later to be adopted and remain in vogue in Japan until the recent introduction of European time. Another noteworthy point of resemblance between the Babylonian and Chinese systems is to be found in the use by each of the sexagesimal cycle, which was also retained in the Japanese calendar.

The duodecimal division of the day was in all probability suggested by the earlier division of the year into twelve moons.

From Babylonia the division of the whole day into twelve parts beginning at sunset passed to Greece, according to Herodotus.[3] Each part was in turn divided into sixty

[1] Guide to the Babylonian and Assyrian Antiquities in the British Museum.

[2] Bigourdan, in the " Annuaire du Bureau des Longitudes," 1914.

[3] Book II, 109 : " *Πόλον καὶ γνώμονα καὶ τᾶ δυώδεκα μέρεα τῆς ἡμερης παρα Βαβυλωνίων ἔμαθον οἱ* "*Ελληνες*," i.e. the Greeks learnt from the Babylonians

subdivisions and these again into sixty. Similar subdivisions of the "hour" were, at any rate in later times, in use in India, having no doubt a common origin in remote ages, long before the dawn of history.

Possibly, the Turanian race, before their westward migration to Mesopotamia, and their neighbours, the Chinese, may in those far-distant ages have evolved a common system of time measurement. There is, at any rate, no positive evidence, so far known, for ascribing to Sumerian origin the system which in historical times had already taken root in China.[1]

The Division of the Day

As by the "natural day" is meant the period between the rising and setting of the sun, and again between sunset and sunrise, it is evident that any equal subdivisions of these periods varied with the season of the year. As soon as the study of the heavenly bodies had developed into an exact science of astronomy it became necessary to adopt a more precise method of dividing the day. Side by side with the natural day with its "temporary" hours, an astronomical day, commencing at midday, was employed. This day was divided into equal parts, or "equinoctial" hours, that is to say, the standard division was based upon the length of day and night at the equinox.

This system was applied in the construction of sundials, which consequently indicate solar or apparent time; but as far back as the fifth century B.C. Meton discovered, by means of the gnomon of the sundial, that the course of the sun in the ecliptic was unequal in its movement, and that the year was not divided into four equal parts at the solstices and equinoxes. In order to rectify this error for purposes of

the bowl and index and the twelve divisions of the day. Πόλος was the word commonly used for the half-sphere, in which the gnomon was set vertically, from its resemblance to the vault of heaven. Such portable sundials, associated with a compass, were in use in Japan until the last century.

[1] I am informed, by Mr. C. Leonard Woolley, that no fresh light has been thrown on the subject of time measurement by the recent excavations at Ur.

exact calculation, a day of average or " mean " length was imagined, with an artificial meridian or noon. The difference between solar time and mean time is known as the " equation of time." We therefore see that what we know as " Greenwich time " is based upon a mean, or average, and that it does not correspond with the solar hours marked on the sundial.

This divergence or error amounts to over 14 minutes about February 11, when the clock is in advance of the sundial, and to over 16 minutes about November 3, when the apparent, or solar time, is in advance of mean time. The days on which the sun and the clock coincide are April 15, June 14, September 1 and December 25. Our forefathers, who were accustomed to reckon by solar time, before the construction of accurate timekeepers became possible with the application of the pendulum to clockwork during the latter part of the seventeenth century, would seem to have found some difficulty in reconciling themselves to the invariable hours; for we find some of the great masters inventing methods for showing both apparent and mean time.

John Smith published his " Horological Disquisitions " in 1694 for the purpose of explaining the difference between apparent and mean time. His book must have supplied a want; for a second edition was necessary in 1708, in the preface to which he states that he has substituted a new table " more easie to bē understood by all Capacities, and of greater Use in the true adjusting of a good Pendulum, and keeping it afterward as near as possible to the Time given by a Sun-Dial." This second table is entitled " The Second Table of Equations, shewing how much a well-adjusted Clock will gain or lose in every Month's Time, if set right to the Sun on the First Day."

Similarly, Dr. Derham, in the Appendix to his " Artificial Clockmaker," published in 1700, wrote:

If you would know whether your Watch [by which he means the going train of a clock] goes well, when kept to go by the

Sundyal if set on *Jan.* 1. it hath gained 23″ you may conclude your Watch goes well. Otherwise you must screw up or let down the *Ball* or *Corrector*, until it loseth, or gaineth according to the Equation Tables.

It was not in fact until long after the adoption of mean time that its substitution for apparent time received official recognition. This first took place in Switzerland in 1780. England followed in 1792, Prussia in 1810 and France not until 1816.[1]

[1] Bigourdan, "Annuaire," *op. cit.* 1914, p. B 8.

CHAPTER I

THE ORIGINS OF CLOCKWORK

NOTHING is definitely known as to the first invention of clockwork, driven by a weight as the motive power. Very elaborate water-clocks would appear to have been devised long before the Christian era, the motions of which involved the use of trains of toothed wheels and pinions. It seems strange that it should not have occurred to any of the inventors of these complicated mechanisms to utilize the force of a falling weight. The force required to raise a bucket of water with a rope and windlass, and the speed with which the bucket would fall, uncoiling the rope from the drum, if the handle were released and there was no ratchet to keep it in position, must have been obvious. Nevertheless, the fact remains that this simple motor was not made use of for keeping a train of clock wheels in motion until the end of the thirteenth or beginning of the fourteenth century.

This is perhaps the more remarkable, because the uncoiling of a rope, wound round a drum, by means of gravitation, and thereby causing a train of wheels to revolve, had been put into operation by Hero of Alexandria, who lived in the latter half of the first century A.D.; whether he was the inventor of the system I am unable to say.

The illustration (Fig. 1) with the Latin text appears in the first printed edition of his "Pneumatics," published at Urbino in 1575. The same design is also reproduced with the text in the original Greek and the Latin translation in parallel columns in Thevenot's folio edition of "Veterum Mathematicorum Opera," published in Paris in 1693. The illustration and the description are of interest, not only on account of the application of a falling weight, but as demonstrating the use of a contrate wheel engaging in a large " lantern " pinion, and of a barrel for winding the rope.[1]

The Latin text may be translated as follows :

Let there be a framework A, B, C, D, containing an arbor E, F, turning easily. To this arbor let there be fixed a wheel

HERONIS.

. Sit theſaurus A B C D, & axis in ipſo diſpoſitus E F facile verſatilis. ei vero cōiungatur rota HK, quam cōuertere oportet. Sint præterea duæ rotæ L M eidem axi intus coniunctæ, quarum L quidē exelictram habeat, M vero radiata ſit. circa exelictram funis inuoluatur. à cuius extremo ſuſpen-

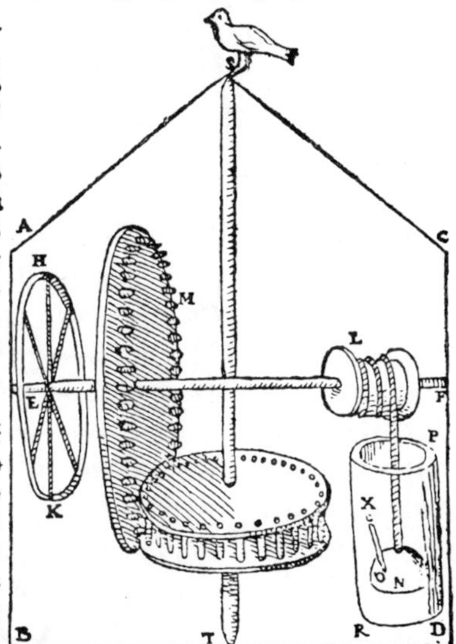

datur clibanus N, tubum habens X O, & in ſummitate ſyringulam, quæ melancoryphi vocem reddat. ſubijciatur autem clibano aquæ vas PR. & à vertice theſauri demittatur axiculus S T, qui facile conuerti poſsit. habeatque ad S melancoryphum, ad T vero tympanum radiatur, cuiùs radij radijs tympani M implicentur. continget igitur conuerſa rota H K, funem circa exelictram inuolui, & clibanum ſuſtinere. at rota dimiſſa clibanum ob grauitatem deſcendens in aquam aere expulſo ſonum edere. ſimul vero & melancoryphum conuerti ob conuerſionem tympanorum.

Fig. 1.—Hero's Piping Bird Mechanism.

H, K, which can be revolved. To the same arbor let there also be attached two wheels L and M within the frame, of which L

[1] The figure is taken, with his permission, from the late M. C. Frémont's " Origine de l'horloge à poids " (Paris, 1915). In neither the Italian nor the French edition does the bird appear, and it was added by M. Frémont to complete the illustration.

consists of a barrel, whilst M is toothed. Let a rope be coiled round the barrel, to the end of which rope is suspended the bell N, fitted with the tube X, O, surmounted by a whistle, to imitate the note of a blackcap. Let the vessel P, R, containing water, be placed below the bell, and from the apex of the framework carry down a stout arbor S, T, which can turn with ease. At the point S fix a bird, and at T a toothed drum, whose teeth engage in the teeth of the wheel M. When the wheel H, K, is turned, the rope will be wound round the barrel and will hold up the bell. But when the wheel is let go, the bell, *descending by its gravity* into the water, drives out the air and gives out the note. At the same time the bird is also made to turn by the revolution of the wheels.

The reason, which doubtless prevented the application of this simple form of power, if the idea had indeed occurred to the inventive mind of the so-called " dark ages," was that the barrel, on which the rope was coiled, would have revolved too rapidly to serve for the purpose of indicating any measure of time. A train of wheels and pinions, governed at the end of the series by a fan or fly, would somewhat retard the fall of the weight through the resistance of the air—a principle afterwards employed for governing the speed of the striking train—but for time measurement the retardation would be practically ineffective ; moreover, the acceleration in the speed of the falling weight would not be impeded. It therefore became necessary to devise some expedient by which the rapidity of revolution could not only be controlled but also equalized.

We may suppose that the solution of this problem was not the result of any sudden inspiration, but that it was the outcome of a long series of experiments and failures.

How and where and when the invention of a contrivance for utilizing this motive force of a suspended weight coiled round a drum, and at the same time adapting it for recording the passage of time, was made will probably remain for ever an unsolved mystery. It is possible that further careful research in the old monastic chronicles may throw fresh light upon the subject.

The monasteries and abbeys in the Middle Ages were the great repositories of knowledge and of the arts and sciences of the time, and it is there no doubt that inventive genius was at work. It is the fact that the earliest weight-clocks, of which we have any definite evidence, were erected in religious houses. It is obvious that a mechanism for sounding a bell automatically to summon the monks to prayer would be a desirable invention ; and it may well be that a wheel, operating an alarum and controlled by a verge and pallets, preceded the invention of the timekeeper, and suggested the application of the same contrivance to the latter.

The perfecting of a controlling device must, undoubtedly, have engaged the attention of the mechanicians of the monasteries for a long period ; and the question why its revelation to the world at large was so long delayed has never been accounted for by all the writers on the subject, who have contented themselves with repeating the same references to early clockwork. We read of St. Gregory in the seventh century ; of Pacificus, Archdeacon of Verona, in the ninth, and of Gebert, afterwards Pope Sylvester II, in the tenth as having devised or set up clocks ; but, whether they were clocks in the modern sense, with a controlling agency, or some form of water-clock is still a matter of uncertainty.[1]

An answer to the question has been propounded by the late M. Charles Frémont, a distinguished French engineer, in his most interesting monograph entitled " Origine de l'horloge à poids," already mentioned.

His attention was directed to the subject by curious drawings contained in an album of architectural sketches,

[1] I use the word " clock " in the general sense of a mechanism for measuring time, although, strictly speaking, it is only applicable from its derivation (L. Lat. *clocca*—a bell) to striking clocks, which were of later date. One of the chief difficulties in determining whether the mechanisms described in the old accounts were weight-driven clocks lies in the fact that every instrument for time measurement, whether a clepsydra, or sundial, or other device is termed " horologium."

the work of one Villard or Wilars, an architect born at Honnecourt, near Cambrai.[1] The date attributed to the album is between 1240 and 1251. The drawings are here reproduced.

The interest of the drawing (Fig. 2) lies in the fact that it illustrates the first instance, so far known, of a contrivance for controlling and regularizing the fall of a weight. The mechanism consists of a four-armed fly-wheel, the shaft or arbor of which is pivoted into two uprights. A third upright between the two carries a small pulley. On the right is a framework, the left upright of which contains the pivot hole of one end of the fly-wheel arbor. The upright on the extreme right carries a second small pulley. Between these two uprights stands a rod pivoted in the bottom of the frame, and passing through the cross-piece at the top, so that it can revolve. On the top of this rod is a spike upon which was fixed the image of an angel with extended forefinger.

M. Frémont explains the operation of the mechanism as follows : a cord, carrying the weight, was passed over the first pulley ; then in front of one arm of the fly-wheel and behind the next. It was then given three turns round the fly-wheel arbor, then one turn round the revolving rod, and

[1] Published with facsimiles by J.-B. A. Lassus in 1858. An English edition with facsimile reproductions of the 33 extant folios of the Album (out of 41 originally), with lengthy descriptive matter, was published at Oxford in 1859 by the Rev. Robert Willis. Attention had, however, been first called to this album by M. Jules Quicherat in the *Revue Archéologique*, vol. vi, 1849. M. Quicherat considered that the mechanism was intended to operate the statue of an angel on the roof of a church. Such an angel in lead existed over the east end of Chartres Cathedral before the fire in 1836. It had an axis, forming part of the casting, suggesting a connexion with a mechanism that had disappeared.

In his " Curiosities of Clocks and Watches " (1866) Wood (p. 18) quotes the indenture made between the Dean and Chapter of St. Paul's Cathedral and Walter the Organer on November 22, 1344, in which occurs this passage : " Le dit Wauter ferra une dyal en Lorloge de mesme leglise od rooss [*wheels*] & totes maneres de ustimentz appartenantz al dit Dyal & au *tourner del Angel* par amunt [*above*] Lorloge " (Brit. Mus. Cart. Cotton, xxi, 24).

I take it that the angel turning with the sun was a device to indicate in dull weather the approximate time of day for the benefit of the illiterate, to whom the figures on a clock dial would be unintelligible.

finally it passed over the last pulley, with a counterweight at the end. The weight caused the fly-wheel to move forward, but at the same time the friction set up by the weight tightened the cord between the two arms of the fly-wheel to such an extent that the wheel was brought up with a jerk and stopped the fall of the weight ; but the impetus given to it was such that it recoiled, and in so doing slackened the coil round the arbor, and in its turn the single

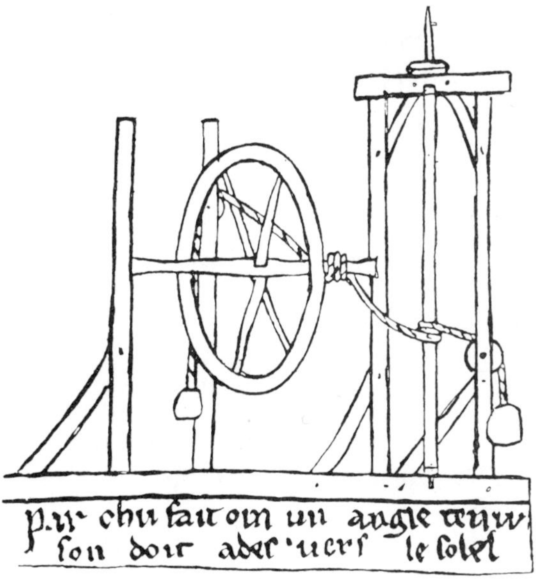

Fig. 2.—Villard's Rope Escapement.

turn round the upright rod. The slackness in the latter was then taken up by the counterweight, which thereby set up a revolving motion in the rod and in the angel fixed on top of it.

M. Frémont points out that, in order to produce the action above described, the cord should be coiled round the arbor of the fly-wheel in the opposite direction from that shown in the drawing.

The inscription in the Picardy dialect reads : " Par chu

fait om un angle tenir son doit ades vers le solel " ; that is : " By this means an angel is made to keep his finger directed towards the sun " (*cf.* footnote on p. 12).

The next drawing, taken from Villard's album (Fig. 3), whilst not illustrating the motive power of a pulley weight, shows the action of a cord coiled round a spindle, with pulley and counterweight, in effecting the partial revolution of an object attached to the spindle.

Here we have the figure of an eagle, with an inscription reading : " Par chu fait om dorner la teste del aquile vers le diachene kant list la vengile ; " that is to say : " By this

Fig. 3.—Villard's Revolving Eagle.

means the head of the eagle is made to turn towards the deacon when he reads the Gospel."

The neck of the eagle, attached to the spindle, is movable ; and, by a light pressure with the book of the Gospels upon the tail of the bird, which would not be perceptible by the congregation, the mechanism was set in action and the head turned round towards the priest.

In all ages thaumaturgy—the working of wonders—has been practised by priestcraft for impressing its miraculous attributes on the minds of the vulgar ; and in the drawings of Villard, M. Frémont believes that an explanation may be found of the question why the secrets of mechanical invention were jealously guarded by the religious houses. If this solution be correct, we may perhaps be justified in supposing that the early clocks of St. Gregory and the rest were in fact driven by a weight with some controlling device for regulating the force exerted by it. But the secret, if secret there was, in regard to the contrivance which finally made timekeeping clocks a possibility, could not be kept for ever ; and, with the end of the thirteenth or beginning of the fourteenth century, they gradually came into general use in religious and afterwards public buildings, to be followed later on by smaller models constructed on the same principle for domestic purposes.

In view of the date of Villard's album it is of interest to note that one of his drawings is an illustration of an elaborate Gothic " clock-house " of many stages and gables which he tells us that he had once seen. If it contained a weight-driven mechanism, it was probably such as would actuate an alarum, for there is no indication of a place for a dial.[1]

I have suggested that a simple contrivance for operating an alarum may have given the idea for a mechanism to control a train of wheels and pinions, calculated in such a way that one of the wheels would revolve once in twelve or twenty-four hours. Such a device was commonly applied to the old " lantern " clocks of the seventeenth and eighteenth centuries.

It consists of a spiked pulley to carry a rope or chain with the weight and counterweight, and furnished with a ratchet and click, so that the weight when pulled up is held in position until the moment for liberating it arrives. This pulley is on the same arbor or axle of a wheel with triangular teeth, set vertically, known as a crown-wheel. A vertical rod or " verge " carrying a hammer at the top is pivoted

[1] The quaint description is as follows :

CEST LI MASONS DON OROLOGE
KI VELT FAIRE LE MAIZON DUNE IERLOGE VESENT CI UNE QUE IO VI UNE FOIS. LI PREMIERZ ESTAGES DE DESOS EST QUARES A .iiij PEIGNONCIAUS [*gables*]. LI ESTAGES DESEURE EST A .viij PENIAUS (?) & PUIS COVERTIC [*Roof*] & PUIS .iiij. PEIGNONCIAUS. ENTRE .ij. PEIGNONS .j. ESPASSE WIT. LI ESTAGES TOS DESEURE SEST QUARES A .iiij. PEIGNONCIAUS. & LI COMBLES A .viij. COSTES. VES ALUEC LE PORTRAIT.

in brackets, or "cocks," just clear of the wheel, and from this rod there project two small pieces of steel, or pallets, at such a distance apart that their centres correspond with the diameter of the crown-wheel. These pallets are set at an angle of about 45° from one another. A pin fixed in the rim of the wheel, resting against a catch-piece at the end of a lever, called the "detent," prevents the weight from acting on the wheel.

The action is as follows : when the given hour arrives the detent is lifted by the motion work in front of the clockwork until it clears the pin. The crown or 'scape-wheel is thus released and the weight is free to fall. But the revolution of the wheel would be too rapid, were it not for the impediment set up by the pallets. The force of the falling weight applied to the wheel causes the tooth at the top to thrust the upper pallet outwards, and the tooth *escapes*, but the swing communicated to the verge brings round the bottom pallet, causing it to lock the tooth below. The impact checks the rotation of the verge, and the recoil causes it to swing backwards. There is thus an alternate *escapement* and locking of the upper and lower pallets, with the result that the T-shaped hammer on the top of the verge oscillates to and fro, its projecting arms striking from side to side on the inside of the bell which surmounts it, until the weight has run down.

Now let us see how this same device may be applied to a series or train of wheels so as to ensure that one of them shall revolve in a given period of time. It is desired that the main or "great wheel" of the train shall complete a revolution in one hour.

Let us imagine the wooden drum of a winch, such as would be used for drawing water from a well, pivoted in a rectangular framework, and having, securely fixed to one end inside the frame, a great wheel. Above the drum and parallel with it would be a second spindle, or "arbor," pivoted into the back of the frame and carrying a pinion, so placed as to engage in the teeth of the great wheel. To the other end of this arbor would be attached the 'scape-wheel. This would be pivoted into a bracket or cock, projecting from the inner side of the front of the frame. In the cock there would be a hole sufficiently large to allow for the free movement of the verge. The verge with its pallets rests below upon a second cock and passes above through a hole in the top plate of the frame. On this plate is mounted a vertical rod bent over at right angles at the top. Half-way down this rod is fixed a horizontal cock, through the pivot hole of which passes the upper end of the verge. The positions of the upper and lower pivot-holes are so measured that the verge is truly perpendicular, and is at such a distance from the 'scape-wheel that the pallets may freely swing to and fro in its teeth, the wheel and pallets together forming the " escapement."

In order to overcome the friction which would arise if the whole weight of the verge were supported by the lower cock, a hole is drilled in the upper end of the verge, and a thin cord, threaded through it, is looped over the returned end of the supporting rod. This cord is made fast in such a way that the lower end of the verge barely touches the lower cock, and that the pallets are centred opposite the teeth of the 'scape-wheel. At the same time the torsion of the cord, as the verge swings to and fro, assists the alternate locking and escapement of the pallets.

So far we have virtually a repetition of the alarum mechanism, except for the interposition of the great wheel and pinion and the absence of the hammer at the top of the alarum verge. It is obvious, however, that without some controlling agency the pallets would be driven to and fro with great rapidity, as in the case of the alarum ; and an instrument for measuring time would not have been achieved. The T-shaped hammer of the alarum probably suggested the solution. Accordingly, a long bar would be forged, having in its centre a boss so drilled as to fit tightly on the

2

verge between the top plate of the frame and the upper cock. Such a bar or balance, or "foliot" as it is called, with its two long arms, would to some extent retard the action of the pallets. But in itself it would be insufficient either to check the rapidity of the motion or to regulate the speed. Doubtless the steelyard, with its graduated arm and sliding weight, suggested the solution of the problem. A number of notches was therefore cut in each of the balance arms, and small looped weights were suspended from them. The proper weight of these riders would be ascertained by experiment, as also of the driving weight attached to the cord on the drum, so as to cause the great wheel to perform one revolution in an hour. The nicer regulation of the motion would then be adjusted by increasing or lessening the leverage of the foliot by suspending the little weights farther from or nearer to its centre.

In explanation of the above somewhat lengthy description of the mechanism I am able, through the courtesy of Mr. Percy Webster, to give an illustration (Fig. 4) of a little fifteenth-century clock in his collection. It will be observed that by that time an improvement had been made by lengthening the train by the interposition of a second wheel and pinion between the great wheel and the 'scape-wheel, thereby producing a slower and steadier revolution of the latter. Side by side with this figure I give for comparison a photograph (Fig. 5) of a little Black Forest clock, dating probably from the end of the seventeenth century, showing that this primitive construction was not yet abandoned. It is entirely of wood except for the pins on the 'scape-wheel, which take the place of the serrated teeth of the crown-wheel.

Indeed, it seems highly improbable that the earliest alarums or going trains of a clock were furnished with a crown-wheel. The man who first conceived and executed one was a genius. Such a departure from all hitherto known forms of wheels must have required profound study and much experiment to secure the proper lock and escape of the pallets ; whereas the pin-wheel was easy of construction and had been in use for centuries, as Hero's mechanism demonstrates (see Fig. 1).

I therefore venture to offer the suggestion that the earliest

Fig. 4.—Fifteenth-century Balance Clock.

Fig. 5.—Black Forest Clock with Foliot Balance.

clocks, or at any rate the earliest alarums, operated with a pin-wheel, and that the crown-wheel was a step in advance of some later date. When that may have taken place it would be impossible to say, seeing that none of the primitive clocks has survived ; and, even if they had, they were subject to constant repair and the substitution of later improvements. The history of De Vick's clock shows that such was the case ; and, when dealing with this subject in Chapter IV, I shall show, from the evidence of Froissart's poem, that there is some ground for supposing that in its original form it worked with a pin 'scape-wheel.

Up to the invention of the first wheel-cutting machine of Dr. Robert Hooke, about the year 1670, all the wheels and pinions had to be cut and filed out of the solid metal by hand, a laborious task, and one not conducing to smooth and accurate running of the trains.[1] For large installations, in which wheelwork was required, a simpler method was adopted in ancient times. This consisted in driving wooden pegs, shaped as teeth, into the periphery of a solid wooden wheel. That this system persisted into the sixteenth century is well shown in the woodcut (Fig. 6), taken from the famous work of Georg Bauer (" Agricola "), the " father of mineralogy," entitled *De re metallica*, 1556. Here may be seen an incomplete wheel with the teeth-pegs lying beside it.

Having, then, succeeded in constructing a train of wheels with a vibrating balance to control them, it still remained to find a means of indicating the twenty-four hours of the day.[2] This was effected in the following way. Instead of merely pivoting the arbor of the great wheel in the front of the frame, it was carried through the frame and terminated in a pinion. This pinion in the primitive clocks took the form of two, three, or more projecting pins, which engaged in the teeth of another large wheel pivoted into the outer framework, and held in place by a cock or bridge. Let us suppose that our pinion had 3 teeth or pins and the so-called " motion-wheel " 72 teeth. Then, as the pinion

[1] Juanelo Turiano (1500–85), clockmaker to the Emperor Charles V, for whom he constructed a wonderful astronomical clock, is stated to have made a machine for tooth cutting.

[2] In the earliest clocks the division of the day into twenty-four hours was adopted, to which reference will be made later.

was caused to revolve one turn by the great wheel of the "going" train in one hour, it would in turn cause the motion-wheel to complete its revolution in twenty-four hours (cf. Fig. 25).

Fig. 6.—Agricola's Water-lifter.

The motion-wheel itself was originally a solid wheel, that is without arms, having painted on it the numbers of the hours, and attached to the framework would be an index, to point the hour as each number in turn came round with the revolving wheel.[1] A revolving hand and a fixed hour circle would be improvements of later invention.

There is, in the Victoria and Albert Museum, in the woodwork department, an exceedingly interesting Italian tarsia panel, a portion of which is illustrated in the Frontispiece to this volume.

Although the panel is attributed to fifteenth-century work, it is evident that the artist was depicting a clock of earlier date, such as I have been describing, as that which might be used in a monastery. This is borne out by the fact that the remainder of the panel represents the utensils used in celebration of the Mass.

Here then we have the solid revolving " dyal " (*diurnal*) wheel with the twenty-four hours painted on it, and the index below. At the back is seen a long verge with an arm at the top, carrying the alarum hammer, and below the crown-wheel. The alarum appears to have been set in motion by hand by means of the crank shown below the dial, and not as a part of the clockwork. Outside the bell an independent clapper is seen suspended, which was used perhaps for tolling the bell at the Elevation of the Host.

Was any striking mechanism combined with the going train of such a primitive timepiece ? In other words, did a fully complete striking-clock spring from the brain of the first inventor ? Such a thing seems hardly possible.

The elaboration of a striking train to work in combination with a going train, and so constructed as to sound the correct number of strokes at each hour, simple as it may seem to-day, must have demanded profound consideration and lengthy experiment. Far more likely is it that the going train was adapted to discharge the alarum work, already devised.

Such a clock exists in the Museum at Nuremberg. A description of it is given in the *Allgemeines Journal der Uhrmacherkunst* of 1884. It is stated to have been placed originally in the belfry of St. Sebald's Church in Nuremberg.

[1] See Chapter IV, on De Vick's Clock.

As may be seen from the illustration (Fig. 7) it operated an alarum, the mechanism of which is placed at the back of the going train. By an ingenious device the alarum-

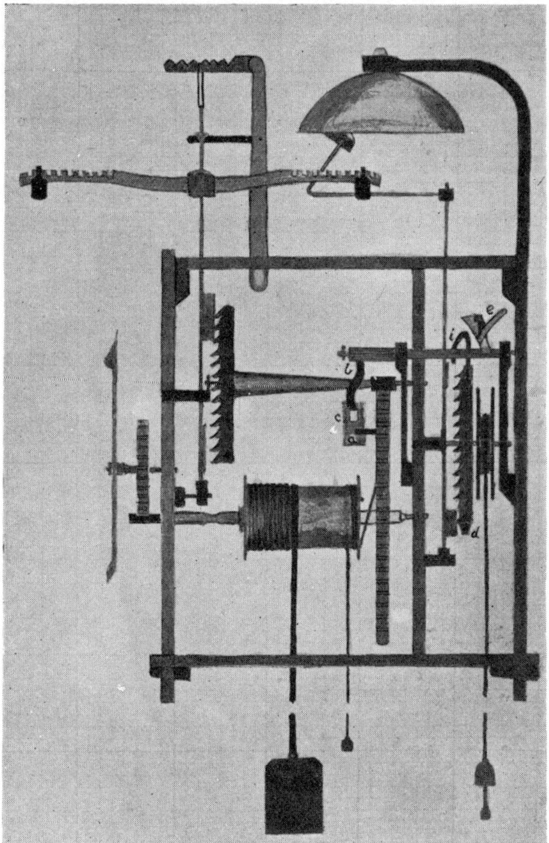

Fig. 7.—Diagram of Nuremberg Clock.

wheel was liberated at the end of every hour, and the bell-hammer was set in motion during one turn of the wheel only. The ringing would, however, last quite long enough to attract the attention of the sacristan or other custodian of the church, who thereupon tolled on the great bell of the church the number of strokes corresponding with the particular hour.

Such I take to have been the function of the earliest clocks. They served not to indicate the passage of time to the public eye, but to give within the monastery a signal at proper intervals for sounding the hour or for ringing the bell to summon the monks to prayer by day and night.

A clock, such as this, by its size (16 inches high) and by its construction was not adapted to work as a turret clock with a dial outside the tower. The hand and motion-wheel, which appear in the illustration, are additions of later date, when the twelve-hour dial was introduced.[1]

The date attributed to this ancient clock is 1400 to 1420, on what ground it is not stated ; but it might well be much older, or else the construction was traditional.

Before leaving it a description of the ingenious mechanism for discharging the warning signal is worth recording.

Fixed to the hour-wheel is a pin *a* which unhooks the lever *b* and thus liberates the alarum-wheel. To this lever is attached with a stop-hinge the plate *c*. On the edge of the crown-wheel is fixed a cone-shaped wedge *d*, which serves as a stop. (For the purpose of illustration it is shown at the bottom of the wheel. When it is at rest it presses against the arm of the lever *e*.) The action is as follows : the pin *a* acting upon *c* raises the lever *b*, causing the arbor to turn and to lift the lever *e*, until it sets free the cone *d*. The crown-wheel at once commences to revolve, and acting on the pallets, sets the bell-hammer in motion. But, inasmuch as the mechanism is not intended as an alarum, but merely as a warning signal, its action must be stopped. Otherwise, as the pin *a* takes some time to pass over *c*, the

[1] The dial of the clock, when it was discovered, had a circle divided into twelve hours. This having been carefully removed, a second twelve-hour circle, painted in red, white and blue, was revealed. The paint of this dial was in turn dissolved, when the original dial, divided in Gothic numerals into *sixteen* hours, came to light.

weight would run down entirely. This is obviated by means of the hook *i*, fixed to the edge of the wheel opposite to the cone *d*. When the wheel has made half a turn the hook raises the arm of the lever *e* a little higher, allowing the hinged plate *c* to drop backwards of its own weight. This liberates the pin *a*, and allows the arm of the lever *e* to return to its place on the edge of the wheel. The cone *d*, at the end of another half revolution, is brought up against this arm, and the ringing ceases until after the lapse of another hour.

The winding arrangement is also worthy of note. Whilst the great weight is descending and unreeling the cord to which it is attached, a smaller cord with counterweight is being wound on the drum at the other end in the reverse sense. When the clock needs rewinding, all that is necessary is to pull the small cord.

The earliest weight-driven time-measurers were doubtless constructed on this plan, a simple-going train actuating an hour signal, without any hand or hour circle. Clocks with an automatic striking train, when such mechanism was first invented, must have been extremely costly, and needed the most skilled craftsmen for their construction.

The town of Montpellier may serve as an instance of the primitive system. One of the regular duties of the keeper of the belfry at Notre Dame was to strike the hours. His deed of appointment constituted him as "Custos campanilis et pulsator horarum noctibus et diebus"—guardian of the belfry and striker of the hours by night and day. This was in 1403; but the high wages that were paid to this officer and his irregularity in striking the time induced the authorities in 1410 to order a large clock from Dijon, which was to have a wooden man, called Jacomart (*anglice* "Jack") to strike the bell.[1]

[1] Renouvier, "Des Maîtres de Pierre de Montpellier," 1844.

CHAPTER II

STRIKING WORK

Before proceeding to discuss the origin of the striking mechanism of clocks a few words must be said as to the division of the day in mediæval times. Before the growth of civic life in power and importance the people necessarily looked for guidance to the religious houses. Monastic life in turn was subject to canonical rule, which prescribed exactly the times for the different offices, the time for the midday rest or siesta (derived from *sexta*), the time for meals, the time for sleep, and so forth. Originally the division consisted of *matins* in the third quarter of the night ; *prima* at sunrise ; *tertia* half-way between sunrise and midday ; *sexta* at midday ; *nona* mid-afternoon ; *vespers* one hour before sunset ; *compline* at the close of the day.

In the course of time Mass was celebrated at the *tertia*, and was immediately followed by the office of *sexta*, after which fast might be broken in the common refectory. *Sexta* thus gradually disappeared as a special midday office, and its place was taken by *nona*, which in turn came to signify midday or *noon*. The change would appear to have been brought about by the necessity for curtailing the period of fasting.

The actual time at which the principal service was held varied in any case, depending upon whether the day was a fast day or not, and upon the rule of the Order ; and it is therefore easy to understand that a clock striking at regular intervals would not fit into the scheme of monastic life.

The beginning and end of the day varied with the season of the year. It was sufficient for the purpose that the clock should show the hours on the " dial " wheel, as will be explained later in the chapter on de Vick's Clock, and be provided with an alarum, to be set at the proper time for rousing the monks for matins. The discharge of the alarum was effected in the following way. Holes were drilled in the dial-wheel at regular intervals near its outer edge. Into the hole corresponding with the time for waking was thrust a pin which projected through the wheel. The dial in its revolution brought the pin into contact with a lever, connected with a detent, which rested against the stop on the alarum-wheel at the back of the clock, and held it fast. The lever as it was raised by the pin in the dial-wheel in turn lifted the detent clear of the stop, and the alarum-wheel was then free to revolve, until the weight which set it in motion had run down.

This setting of the alarum was the duty of the sacristan, and was termed *horologium temperare* or *dirigere*. In the Rules of the Cistercian Order, for instance, it is laid down that the sacristan shall set the clock and make it sound, whereupon he was to rise and light up the dormitory and the church. It is true that this Ordinance dates from 1119, and applied to a water-clock ; but it would equally apply when weight-driven clocks were introduced. The ringing of the alarum by means of the falling weight is expressed as *horologium cadit*.

This method of operating the alarum was generally employed in domestic clocks down to the seventeenth century, and was perpetuated in Japan until European clocks came into use.

The canonical hours were based upon the primitive sundial with a staff at right angles to the dial, so that the length of the hours constantly varied. It is easy to see that such a system would not accord with an instrument whose hours were all of equal duration, even if their length for the day and night respectively varied according to the season of the year.

In any case the cloister bell was rung as a signal to the monks to call them to the performance of their various duties ; and a clock which struck the hours at fixed intervals, not necessarily coinciding with the times prescribed by the

Rule, would only tend to confusion. The same objection would apply to church clocks, where such existed.

The reasonable conclusion is that the automatic striking mechanism was an improvement dictated by the necessities of civic life. As towns grew in importance, rules for their good governance were formulated, to which the whole body of the citizens had to conform. Thus a bell would be rung to announce the opening and the closing of the gates. Again, it rang to tell the hours for commencing and ending work, and for the midday meal and rest. The curfew sounded for putting out fires and lights. Another bell, called the first watch, at dusk, and the second watch at midnight, and so on. Supposing that these bells were rung at stated intervals, varying only with the season, a mechanism which would not only give the signal, but also sound the actual hour, would be an invaluable acquisition.

The time had come when the antiquated rules of the monastery, founded on ancient Jewish and Roman custom, were no longer suited to the needs of busy commercial and industrial centres, and the alarum-clock of the convent was gradually replaced by the striking-clock of the civil community. But that is not to say that a religious edifice would not have been employed for the purpose. In the absence of some large public building or palace, the tower of the parish church or the belfry of a cathedral would then naturally afford the most suitable position for installing a striking-clock, with some form of external dial, visible to the citizens, and for hanging a suitable bell, which could be heard far and wide.

According to Bilfinger [1] the town paid a contribution for the right to use the tower of the church, and made itself responsible for the maintenance of the clock and the parish bells.

Another reason for believing that the striking mechanism

[1] Gustav Bilfinger, "Die mittelalterlichen Horen und die modernen Stunden," Stuttgart, 1892.

was not of monastic invention consists in the fact that, so far as positive evidence goes, the earliest striking-clocks were installed by laymen.

As with the weight-clock with its verge escapement, so with the striking mechanism, there is unfortunately no record as to when, where, or by whom it was first given to the world, and its origin is likely to remain an unsolved problem.

The idea of sounding the hours by means of " clockwork " was in itself nothing new ; but these mechanisms of Oriental origin were driven by water-power. Some of them included complicated astronomical movements, which may account for the fact that such additions quickly followed the introduction of the striking work.

It must not be supposed that the new invention spread rapidly. The construction was tedious and costly, and only rich princes and wealthy cities could afford the expense. Even Paris, as we shall see, had to wait until 1370 for its first striking-clock. But, for many long years after that, the smaller communities had to remain satisfied with a simple clock that showed the time, the hours being tolled by hand on a bell, hung either in some public building, or in a church tower, hired by the community for civic use.

The oft-quoted passage from Dante's " Paradiso," Canto X, line 139, whilst it affords no proof that he had in mind anything more than the signal for ringing the bell for matins, does not necessarily exclude the idea of a striking-clock which was also furnished with alarum work. The lines are as follows :

> Indi, come orologio, che ne chiami
> Nell'ora che la sposa di Dio surge
> A mattinar lo sposo perchè l'ami,
> Che l'una parte l'altra tira ed urge,
> Tin tin sonando con sì dolce nota
> Che 'l ben disposto spirto d'amor turge ;
> Così vid'io la gloriosa ruota
> Muoversi, e render voce a voce, etc.

In a note to this passage in Buttura's edition of the Poem,[1] the editor explains that by "orologio" is meant the "svegliarino," i.e. alarum, and that the words "Che l'una parte l'altra tira" are intended to express the method by which the mechanism of the clock causes and prolongs the sound of the bells. Fraticelli,[2] in his Commentary, also considers that Dante is speaking of an alarum.

In fact, the words "tin tin sonando" suggest the to-and-fro action of the alarum hammer rather than the striking of a few blows on a single bell. This seems to be confirmed by the allegory of the glorious wheel in motion which gives forth voice to answering voice.[3]

The other reference to clocks by Dante occurs in Canto XXIV, line 13, as follows:

> E come cerchi in tempra d'oriuoli
> Si giran sì, che 'l primo, a chi pon mente,
> Quïeto pare, e l'ultimo che voli.

That is to say, that anyone who pays attention to the wheels composing a clock will observe that they revolve in such a way that the first one appears to be still, whilst the last one flies.

These lines might be taken to refer to the striking train of a clock, with the slow movement of the great wheel, and the rapid revolution of the fly. The "Paradiso" was composed between 1317 and 1320; so that, if this assumption be correct,

[1] Paris, 1823. [2] Florence, 1886.

[3] An amusing incident occurred in connexion with Sir Henry Irving's production of Sardou's play "Dante" in 1903. The action is supposed to take place between 1303 and 1314. There is a thrilling scene when the wicked Cardinal Colonna is to die on the *stroke* of midnight. On the wall was the *dial* of a clock. Beneath it was a door, which was flung open, showing the *pendulum* swinging to and fro, whilst the *minute hand* moved forward ineluctably towards the fatal hour. I ventured to point out to Sir Henry that this was a gross anachronism, and that most of the details were some 200 to 350 years antedated, and received in reply the following letter: "*May* 11, 1903. I am much obliged for your interesting letter. It is all quite true. The whole history of mediæval clockwork is, as you say, obscure. For that very reason the precise date of the clockwork in 'Dante' is of no great consequence. When the period of a play is a remote century no dramatist ever sticks at anachronisms. Witness Shakespeare's 'Cleopatra's billiards'!"

Dante must have seen a striking-clock during his many wanderings before the Visconti clock of Milan was constructed. But it can hardly be supposed that in symbolizing a clock for the purpose of his allegory he was thinking of two distinct and dissimilar mechanisms. On the other hand, it seems more likely that he was comparing the slow movement of the going train with the swift revolution of the released alarum-wheel.

Prof. Belgrano[1] quotes the manuscript Chronicle of Galvano Fiamma as stating under the year 1309 that at this time an iron clock was set up in the Sant' Eustorgio in Milan; and Sig.ʳ Caffi[2] relates that this same clock was restored in 1333, again in 1555, and "renovated" by a Dominican friar in 1572. This, so far as I am aware, is the first record of a public or semi-public clock in Italy. We know that Dante returned to Italy from Paris in 1311 to pay homage to Henry VII on his coronation as Roman Emperor in Milan, and it is therefore conceivable that it was this clock that inspired Dante's imagery. But we may fairly argue that it was not a striking-clock, when we contrast this meagre notice with Fiamma's eulogy of the striking mechanism of the clock of 1335—*horologium admirabile*—in the glowing terms quoted later by Muratori.

The records of the clock of St. Paul's Cathedral in London show that the Chapter paid for the services of a clock-tender as early as 1286; whilst a "great new clock" was set up in Canterbury Cathedral in 1292. Again, a public clock-tower and "horologium" are alleged to have been erected at Westminster in 1288 out of a fine imposed for corruption on Judge Ralph de Hengham. The whole story is based on tradition. But there is no evidence to show that any of these clocks gave warning of the hour without the intervention of a bellringer.

The "horologium" made by Ricardus de Wallingford for

[1] L. T. Belgrano: "Degli antichi orologi pubbici d'Italia" in the *Archivo Storico Italiano*, Series 3, vol. vii, Florence, 1868.

[2] Caffi, "Della Chiesa di Sant' Eustorgio," p. 193, Milan, 1841.

St. Albans Abbey would appear, from the illustration in the "Lives of Benefactors" (Cotton MSS. Nero, d. 7, in the British Museum), to have been an elaborate planetarium, and certainly not a turret clock. At the centenary celebrations at St. Albans in 1926 the mechanism was so described. In any case it was unfinished at the abbot's death in 1334.

Moreover, if we may assume that striking-clocks were a secular invention for secular needs, we should hardly look to ecclesiastical buildings for their *first* appearance.

Italy was at that time in advance of other European countries in all the arts and sciences; and the fact that the earliest authentic striking-clocks were of Italian origin confirms Bilfinger's opinion that "of all European countries Italy has the best right to claim the invention" (*op. cit.*, p. 175). He considers that the first real striking-clock was that which was erected in the Palace Chapel of the Visconti in Milan in 1335.

Fiamma's description of the clock, as quoted by Muratori, is as follows:

> Et licet in summitate campanilis sint plures campanæ, est ibi unum horologium admirabile, quia est unum tintinabulum grossum valde, quod percutit unam campanam XXIV vicibus, secundum numerum XXIV horarum diei et noctis, ita quod in prima hora noctis dat unum tonum, in secunda duos ictus, in tertia tres et in quarta quatuor, et sic distinguit horas ab horis, quod est summe necessarium pro omni statu hominum.

Which may be translated as follows:

> Moreover, in the top of the belfry are several bells, and in it there is a remarkable clock, because it has a very heavy hammer which strikes one of the bells at twenty-four intervals in accordance with the number of the twenty-four hours of day and night; that is to say, it gives a single sound at the first hour of the night, at the second two blows, at the third three and at the fourth four, and thus it distinguishes hour from hour, which is of great necessity for all conditions of men.

It was shortly after the Milan clock that a striking-clock, designed by the celebrated astronomer and physician, Jacopo Dondi, was erected in 1344 in the entrance tower of the Carrara Palace at Padua by Prince Ubertino. It showed the phases of the moon, the signs of the zodiac, and possessed other astronomical features. Unfortunately it perished in an assault on Padua by the Milanese in 1390, but such was its fame that the surname of " del Orologio " was conferred on its author and has descended in the family ever since. Jacopo died in 1359, and his Latin epitaph in the Cathedral of Padua claims for him the invention of the clock.

This clock is spoken of by Muratori[1] as " Horologium, quo per diem et noctem quatuor et viginti horarum spatia sponte sua designarentur, in summa turri constituendum curavit," i.e. " He caused to be erected in the top of the tower a clock, which *of its own accord* indicated the intervals of the four and twenty hours by day and night."

Clearly, the automatic striking of the clock is here intended ; because the rotation of the dial, or the movement of a hand, would not be visible during the hours of darkness.

It must here be mentioned that confusion has arisen as between the clock *designed* by Jacopo, and executed by a clever workman, Antonio of Padua, and the wonderful astronomical clock " made with his own hands " by his son Giovanni. This clock was begun in 1348 and completed in 1364. Some years later he sold it to Visconti of Milan, who installed it in the Castle at Pavia. The clock appears to have been constructed on the lines of Jacopo's, to which Giovanni added his own marvellous planetarium.[2]

Philippe de Maisières, a contemporary and friend of Giovanni Dondi, in his " Songe du vieil Pélérin," written *ca.* 1385, and quoted by Dubois,[3] gives a description of the

[1] Muratori, " Annales Mediolanenses," Book XVI, p. 171.

[2] A long article by Prof. A. Gloria on the two clocks appeared in 1896 in vol. 54, pp. 675–736 of the *Atti del R. Istituto Veneto di Scienze.*

[3] Pierre Dubois, " Collection archéologique du Prince Soltykoff," p. 19, Paris, 1858.

instrument made by Dondi which was "by some called a 'sphere' or clock to show the movements of the heavenly bodies." He says that it contained—

> all the movements of the signs and planets with their cycles and epicycles and differences by multiplication, wheels innumerable with all their parts, and each individual planet in the said sphere. Any night could be clearly seen the position of the planets and the stars, in what sign and degree; and so subtly is this sphere constructed, that, notwithstanding the multitude of wheels that could not be counted accurately without taking down the mechanism, its whole motion is governed by one single weight.

He adds that Dondi forged the clock from brass and copper *with his own hands*, and devoted the whole of sixteen years to its construction, and that "solemn astronomers" came from far-off countries to pay reverence to the master and his work.

Genoa received a striking-clock in 1353, and Bologna in 1356.

On the other hand there is some other evidence of automatic striking-clocks of earlier date than that of Milan. A clock was erected at Caen upon the bridge, and its bell in 1314 bore the following inscription:

> Puisque la ville me loge
> Sur ce pont pour servir l'orloge,
> Je ferai les heures ouir
> Pour le commun peuple rejouir.

But even this is not conclusive. The bell "served" the clock; but it does not follow that it was connected with the mechanism.

There is a record in the Archives of the Vatican that a clock-house was built in the Palace of the Popes at Avignon in 1353. It is hardly likely that a special building would have been constructed to contain anything but a striking-clock, when the new invention had already been installed eighteen years earlier in Milan, and Jacopo de Dondi had completed his clock at Padua in 1344. In 1375, 100 francs was paid to Pierre de Sainte Béate for his expenses for travelling from Paris to Avignon and his sojourn there and for constructing a clock; but it cannot have been a "turret" clock for so small a recompense.

Whether these were the earliest self-striking-clocks or not, it is certain that from this time onwards such clocks were installed in many of the principal cities of the Continent, whilst many of them were further embellished by more or less elaborate astronomical and mechanical motion work.

Probably the first striking-clock erected in England was that ordered by Edward III for the Palace of Westminster. Wood (p. 25) states, but without quoting his authority, that the King caused a clock tower of stone to be erected about the year 1365–6, opposite the Hall, and that it "contained a clock, which struck every hour on a great bell, to be heard into the hall in sitting time of the Courts, and the same clock in a calm night might be heard in the City of London."

Now Rymer, in "Fœdera," vol. vi, p. 590, gives the text of the patent[1] or passport granted in 1368 to three Dutch clockmakers by name John Unenam, William Uneman (*sic*) and John Lietuyt of Delft, permitting their sojourn in England for the purpose of exercising their craft.

The reasonable inference is that these men were invited to England, and placed under the King's own protection, for the purpose of constructing the clock and installing it in the tower which had just been built. In confirmation of this supposition we have the entry, by the Monk of Malmesbury, under the year 1373, "hoc anno horologia distinguentia 24 horas primo inventa sunt." (In this year clocks distinguishing the twenty-four hours were first invented.)[2] The worthy chronicler was doubtless ignorant of what had already taken place on the Continent, and supposed that this was the first clock of its kind. It may

[1] Quoted by Wood, *op. cit.*, p. 51.

[2] Eulogium historiarum a Monacho quodam Malmesburiensi in vol. 9 of "Scriptores rerum britannicarum."

well be that the Monk, whose Chronicle extends to 1413, was not quite correct as to his date, because Wood (p. 26) quotes a record in an Issue Roll of 1371 of two pounds paid "to John Nicole, Keeper of the great clock of the Lord the King within the Palace at Westminster, taking per day sixpence for his wages for the custody of the clock aforesaid."

The system originally adopted for striking the hours was that which prevailed in Italy, and became known as Italian time. Following Jewish and canonical practice, the day ended and commenced at dusk; that is to say, about half an hour after sunset. The day was divided into twenty-four hours, its completion being marked by twenty-four strokes on the bell. The clocks of Milan, Padua and of other Italian cities were, as the descriptions show, constructed on this principle. The same practice was followed when the striking mechanism came into use in other parts of Europe. De Vick's clock, as described by Froissart, was made to strike up to twenty-four (see Chapter IV), and the same may be said of Edward III's clock at Westminster, if we may rely upon the statement of the Monk of Malmesbury. The Italian tarsia panel in the South Kensington Museum (*see* Frontispiece) also depicts the twenty-four-hour dial.

The reason for adopting dusk as the moment for the stroke of twenty-four was that all work ceased as soon as darkness set in. But in practice it led to great irregularity, because the actual moment depended, not upon the mechanism of the clock, but upon the caprice of the clock-tender. Not only was sunset an ever-changing factor according to the time of the year, but the period of twilight after sundown varied as the sky was clear or overcast.

Strange as it may seem, this system held its ground in Italy even into the nineteenth century, and remained in use in certain Eastern countries north of the Alps during the eighteenth century, when it was commonly called "Bohemian" time.[1]

The inconvenience of counting the strokes with such high numbers must soon have made itself felt; whilst the enormous waste of power in the striking train necessitated constant rewinding. These considerations led to the adoption of the ancient division of the day into two series of twelve hours each, with a corresponding alteration in the dial, and also, no doubt, of the hour circle and the revolving hand. Nevertheless, in Italy at any rate, the custom of counting up to twenty-four persisted, the number of the hour following the first series of twelve being added to the latter figure, a system that has in recent times been adopted in continental railway time-tables, and in telegraphy.

Where and at what date the system of reckoning twelve hours from midnight and again twelve hours from noon, with a dial to correspond, was first introduced is a matter of uncertainty; but it may undoubtedly be ascribed to the latter part of the fourteenth century. Bilfinger (*op. cit.*) gives a lengthy list of the earliest striking-clocks set up in Germany, France, Switzerland, the Netherlands and England, with numerous references in the archives to the time of day; but so many of these refer to hours between midnight and noon that, if the day was reckoned from midnight, one cannot say positively that the clocks were constructed to strike twice twelve.

The earliest reference in England to an hour after noon, reckoning from twelve, appears in the State Memorandum quoted in Rymer's "Fœdera" (vol. vii, p. 151), which chronicles the death of Edward III on June 21, 1377. It took place "circa horam septimam post horam nonam," i.e. "about the seventh hour after noon." Bilfinger says that this must refer to a private *striking*-clock of the King's erected at Shene; surely a gratuitous assumption which the

[1] Goethe, in his *Italienische Reise* (1816), tells of the impossibility of converting Roman time into mean time.

mere mention of the time of day does not warrant. It is more natural to suppose that the hour, *about* 7 p.m., was roughly determined by the position of the setting of the sun, which would have taken place about 8.15 p.m. on June 21 O.S., or it may have been derived from a sundial.

Certainly the double twelve-hour system had taken root in England some time before the close of the fourteenth century. Thus Chaucer, in his " Treatise on the Astrolabe," written in 1391, part II, ch. iii, says : " I wot wel that in oure orisonte from 11 of the clokke un-to oon of the clokke . . . hit is hard to knowe. I mene, from 11 of the clokke biforne the houre of noon til oon of the clokke next folwing." Again, in the " Canterbury Tales " in " The Parson's Prologue," lines 2–5, we read :

> The sonne fro the south lyne was descended
> So lowe that he nas nat, to my sighte,
> Degreës nyne and twenty as in highte.
> Foure of the clokke it was tho, as I gesse.

I find in Rymer's " Fœdera " occasional " Memoranda " relating to the use of the Great Seal. The hour is carefully noted when it was affixed to documents ; when it was delivered to the King, or by him returned to its custodian, the time recorded being most frequently *post* or *circa horam nonam*. This phrase occurs in 1341 and subsequently, and evidently means not 9 a.m. but the hour of noon, if we compare the same words in the record of Edward III's death. Other entries, such as " hora prima," " hora tertiarum," " hora vesperarum "—this latter occurring as late as 1383—seem to indicate that the canonical hours—prime and terce and vespers—were still in use, and give no clear indication of hours " of the clock." But in 1401 and 1407 we get the expression " circa horam decimam ante prandium " (ten o'clock before dinner), which clearly shows that with the beginning of the fifteenth century the new system had taken root. Finally, the Memorandum which chronicles the death of Henry V in 1422 ("Fœdera," vol. ix, p. 253) states that it took place between 2 and 3 a.m., "inter secundam et tertiam horam post mediam noctis."

As regards the Continent, amongst the archives quoted by Bilfinger, the earliest reference to the 2 × 12 hour system and an hour after noon appears to occur in those of Lübeck in 1380 : "Veer stunde na mitdage" (four o'clock after mid-day).

At Bruges we have, on May 3, 1382 : "Snavens naer den neghen ueren in de duysterheid" (in the evening after nine o'clock in the dusk).

In "Chronicles of the Netherlands" it is recorded that a conflagration occurred in Tournai in 1384 "par ung vendredi de nuit environ II (deux) heures après XII" ; but here it is possible that the record was made at a later date.

A record of Cologne of the year 1385 speaks of "zo eynre urin na middage" (about one o'clock after noon).

A chronicler of Freiburg in Switzerland speaks of an event happening on September 19, 1386, "in secunda hora post mediam noctem" ; but here again the entry may not have been made at the time.

But when we come to Froissart we are on surer ground. In his description of important events, and even of more ordinary occurrences, he is most precise in stating the time at which they happened. The first portion of his Chronicles ends with September 1377, and up to that date he makes use of the canonical hours. The second portion covers the period from 1377 to 1386, and is principally concerned with events in Flanders, where the introduction of public clocks had perhaps become more general than elsewhere north of the Alps. The municipal accounts of Ghent disclose the expenditure incurred in the installation of a striking-clock in the public belfry between the years 1370 and 1378. At the end of 1379 disturbances broke out in the city, and Froissart recounts how the Count of Flanders was expected

" à VIII heures," and " le matin à VIII heures il vinrent." Again, the joyful news of the victory at Beverhoutsveld reached Ghent " le dimanche au matin à VII heures."

In Paris too, on the occasion of unrest in 1382, it was decreed that no one should be found abroad under pain of death " apriès le *son* de IX heures."

In the last instalment of the Chronicles, dealing with France between 1389 and 1400, the references to the modern time system are of frequent occurrence.

To sum up then, we may conclude that in North-West Europe striking-clocks began to be set up in buildings under lay control from about the year 1370 onwards, that these clocks were fitted with a twelve-hour dial, and that this arrangement brought with it the modern usage in civic life of reckoning time from midnight and midday, with final abandonment of the rule of the monastic Orders.

FIFTEENTH-CENTURY CHAMBER CLOCK

By courtesy of Mr. Percy Webster.

CHAPTER III

DOMESTIC OR CHAMBER CLOCKS

So far we have been concerned solely with "turret" clocks, that is, such as were installed in monasteries and the towers and belfries of public buildings and churches.

By "domestic" clocks is meant clocks of smaller dimensions, intended for private use. Here again we have no certainty as to when they were first made, nor whether their construction was attempted prior to the invention of the striking mechanism.

The earliest reference to a chamber clock with which I am acquainted occurs in the "Roman de la Rose," in the second portion of the poem, written by Clopinel, commonly known as Jehan de Meung.

From the internal evidence of the poem M. Pierre Marteau, in his edition published in Paris in 1879, comes definitely to the conclusion that the work was completed before 1282. M. Guillon,[1] on the other hand, assigns the date to the last five years of the thirteenth century, a difference at the most of about twenty years.

The passage in question (lines 21815–18) describes how Pygmalion has completed his statue in ivory, but cannot call it to life. He arrays the image in wonderful vestments and jewels ; but all in vain. He then sings love songs and plays on various instruments—

> Et refait sonner ses orloges
> Par ses sales et par ses loges
> A roës trop sotivement (très habilement)
> De pardurable movement (à mouvement continu).

These lines have been Englished by Mr. F. S. Ellis, in his translation of the "Roman de la Rose" in the "Temple Classics" (1900), as follows :

> Then of clocks
> The tongues he skilfully unlocks

[1] "Jean Clopinel," Paris, 1903.

With subtle wheels arranged that so
They may in halls and chambers go
With tireless motion.

If there were domestic striking-clocks at such an early date the statement would be very remarkable. But here we are confronted with the same difficulty in regard to the word "orloges." They may quite well have been, and probably were, water-clocks, whilst the continuous ringing suggests that some sort of alarum was set going.

References occur in the Archives of the Vatican to chamber clocks provided for the Palace of the Popes at Avignon. The earliest is one mentioned as being in the Pope's *garderobe* in 1343. In 1363 twenty florins were paid to Jean de Venise for a clock for the Pope's *arrière chambre*.[1]

M. Havard, in his "Dictionnaire de l'Ameublement" (vol. ii, p. 1172), mentions that a clock was made for John, King of France, in 1359, whilst he was held a prisoner in London. The price paid for it was seven nobles (£2 6*s*. 8*d*.).

In the "Comptes des Batiments Royaux" of Paris for 1365 entry is made of "un estuy de bois d'illande pour hebergier l'orloge de M. le Dauphin, qui sonne les heures audit Louvre."[2]

I have mentioned in Chapter IV Pierre de Sainte Béate as clock-maker to Charles V, who in an Act of October 28, 1377, is described as "nostre amè orlogeur." In that year he was paid 200 gold francs for a clock in the King's Hotel at Beauté-sur-Marne, 65 francs being paid to Jehan Jouvence for its bell. That was probably a turret clock; but in 1377 he also made two small clocks for the King. In an Act of November 24 Charles disburses "VI^{xx} frans d'or pour paier un *orloge portative* que nous avons acheté de Pierre de Saincte Beate, nostre orlogeur." Again, on December 23, 1377, he pays "a maistre Pierre l'orlogeur XX frans pour un petit horloge qu'il a fait pour nous."

[1] R. Michel, in "École Française de Rome: Mélange d'archéologie et d'histoire, 1905: Les premières horloges du Palais Pontifical d'Avignon."

[2] Alexandre de Laborde, "Notice des Emaux du Louvre," etc.

On March 18, 1378, there is an entry for " 140 frans pour un autre orloge." [1]

Much older than any of the above house clocks is " un reloge d'argent tout entierement sans fer qui fut du Roy Philippe-le-Bel avec deux contrepoids d'argent emplis de plon." Philip reigned from 1285 to 1314. The entry occurs in an inventory of 1380 ; so that, if the record is correct, this is perhaps the oldest chamber clock of which we have any certain knowledge. The fact that it had two weights cannot be accepted as proof that it was a striking-clock, for the second weight was probably used for operating an alarum.

It is noteworthy that all the above clocks are of French origin. That Italy with its splendid public clocks should have produced no chamber clocks in the fourteenth century seems improbable ; but I have not found records of any.

During the fifteenth century portable, or semi-portable, clocks became comparatively common ; but as it is my purpose to deal with the evolution of clockwork, I shall not attempt to catalogue the examples of which records or illustrations have come down to us. I shall only mention that a clock with its wheels and motions formed a favourite subject for a homily in that century, under the title of " L'Orloge de Sapiense," founded on texts from the Wisdom of Solomon.

M. Planchon [2] illustrates two such clocks from manuscripts in the Bibliothèque Nationale, but these represent clocks of very large dimensions, and their design is probably fanciful. The manuscripts are attributed to the fifteenth century.

The British Museum also possesses two illuminated manuscripts in which such clocks are depicted. One of these (MS. 15288) ascribed to Jehan Soulchen, a German Jacobin, and dating from about the middle of the fifteenth century, contains a large miniature, occupying half of the

[1] Michel, *op. cit.*
[2] " L'Evolution du mécanisme de l'horlogerie," Paris, 1918.

first page. It illustrates a chapel with a preacher in a pulpit, and monks and others below. On the floor is seen an oblong portable clock standing on a Gothic stool, through a hole in which the cords and weights hang down. The clock itself is richly decorated with crocketed spires, and has an open-work dome, in which are the bell and hammer. The dial is numbered I to XII, and there is a hand.[1]

The other clock of Sapience occurs in the works of Christine de Pisan (Harleian MSS. 4431), on folio 96, a French manuscript assigned to the first half of the fifteenth century. The illustration shows clockwork, raised on a pillar, with a bell and weights, but without a dial. Three women sit below, whilst a fourth, as an angel, hovers above and with outstretched finger points to the wheels. As an example of the sort of exhortation suggested by the clock, I transcribe the passage which follows the words of Solomon :

> L'attrempance estoit aussi appellee deese et pour a que nostre corps humain est compose de diverses choses et doit estre attrempe selon raison peut estre figure alarge (horloge) qui a plusieurs roes et mesures et toute fois ne vault rien lorloge sil nest attrempe semblablement non fait nostre corps humain se attrempance ne sordonne.

It must be understood that all these early domestic clocks were driven by weights, and that as a consequence they were hung against a wall, or were supported on some more or less elaborate structure sufficiently high to permit of the descent of the weights.

The illustration which faces this chapter gives an excellent representation of the primitive work of the early domestic clock. It is the movement, stripped of its outer casing and bell, of a clock in the collection of Mr. Percy Webster, to whose courtesy I owe the reproduction. It is probably of Italian workmanship, dating from the fifteenth century.

Until the application of the pendulum as a regulator in 1657 the structure of these weight-driven clocks remained

[1] An outline drawing is given in vol. 34 of *Archæologia*.

virtually unaltered, as will be seen from the illustration of a chamber clock in Capriglia's "Misura del Tempo," written in 1663. (Fig. 8.)

It will be noticed that, in place of the bar balance, a

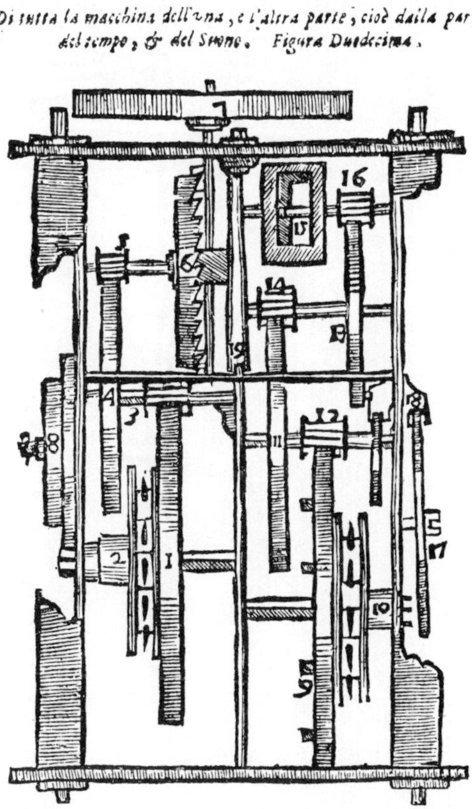

Fig. 8.—Capriglia's Chamber Clock.

circular balance is used, and that there are three wheels in the going train instead of only two as in his turret clock (see Fig. 9). This form of escapement was used in our own early lantern clocks (see Fig. 19), and was perpetuated in the wooden clocks of the Black Forest long after the pendulum had superseded the balance.

It is not until the force exerted by a coiled spring was applied to clockwork that truly portable clocks, and eventually watches, became a possibility. The use of a spring motor is attributed to the beginning of the fifteenth century ; whilst the first small portable clocks are ascribed to the Nuremberg smith, Peter Henlein, about 1511.

CHAPTER IV

HENRY DE VICK'S CLOCK

It has been so much the custom of writers on horology to refer to this clock and to describe and illustrate the mechanism of its verge escapement, surmounted by the foliot balance as a regulator, as to create the impression that it was the prototype of mediæval clockwork. The reason is not far to seek. We know that, in point of time, many more important clocks had preceded it ; but no exact description of their mechanism has come down to us. Such is not the case with Henry de Vick's clock ; for what purported to be a minute account of it by the celebrated clockmaker, Julien Le Roy, was found amongst his papers after his death. From this account the illustrations which have been copied by one writer after another have been derived.[1]

The history of the original clock may be told in a few words. Already many towns in Italy, in France, in Germany and in England had been provided with clocks, but Paris, the capital of France, and the centre of its artistic and scientific life, still lagged behind. To remedy this defect was the ambition of Charles V, surnamed " The Wise." Unfortunately, for his purpose there appears at the moment to have been no native artificer capable of executing the

[1] Le Roy's manuscript was first published by Louis Moinet in chapter i of his " Nouveau Traité Général . . . d'Horlogerie," published in Paris in two volumes in 1848 with an elaborate plate of illustrations. The trains as given by Le Roy when he saw the clock (? 1731) are as follows :

Going train.	*Striking train.*
Great wheel, 64—8.	Great wheel, 64—8.
Second wheel, 60—8.	Second wheel, 64—8.
'Scape-wheel, 30 (*sic*).	
Foliot balance.	

Motion work.—Pinion of 8 engaging in hour wheel of 96 carrying 12 pins to raise the detent.

Driving weights.—500 lb. on each train.

This is the clock described and illustrated in Britten's " Old Clocks and Watches," and in most other horological works, and as lately as in 1924 in " Time Measurement," by L. Bolton.

work, namely, a clock that would automatically strike the hours, to be erected in the Royal Palace, now the Palais de Justice.

The King was therefore obliged to look abroad, and his choice fell upon a German, called after his birthplace, Heinrich von Wiek. Whether this place was situated in Würtemberg, as some suppose, or near Duurstede on the Rhine, or in Lorraine, or elsewhere, is uncertain. In France, at any rate, he was known as Henri de Vic.

Henry de Vick—as following English custom, I shall call him—accepted the King's invitation ; and it is recorded that he received for his services six Paris sous per day, with free quarters in the tower of the palace.

The work was commenced in 1362. Simple as the machine was, as compared with modern mechanisms, it is not surprising that the work occupied a number of years. It was the work of the blacksmith and the locksmith combined. Every piece of the framework had to be forged, every wheel cut out truly circular, every tooth laboriously formed with the file. One may suppose that the execution of each piece was not always successful, and that the work had to be done over again. The true adjustment of the going train with the striking train, with the lifting pieces and the locking plate would present innumerable difficulties.

What previous experience de Vick may have had is unknown, or whether he was acquainted with the clocks already made in Italy ; but, as I have mentioned in a previous chapter, automatic striking work was a comparatively recent invention, and each artificer would wish to exercise his own ingenuity in working out the detail of each mechanism that he was called upon to construct. Much experiment was doubtless then involved, as in our day with the motor-car, the aeroplane and the like.

The work was finally completed in 1370. Meantime, the celebrated bell-founder, Jean Jouvence, had been entrusted with the casting of the bell upon which the hammer was to strike the hours ; and this was successfully hung in the upper part of the tower in the following year.

A totally false impression of the nature of this primitive clock has arisen, and has been perpetuated from the description of Julien Le Roy, previously mentioned. Julien Le Roy died in 1759, and his account was written at least 350 years after the completion of de Vick's work. It is not to be supposed that in all those years the clock in the Palais had not been repaired again and again, and remodelled to conform with such improvements as later progress in the art suggested. We know that such was the case ; because, under Charles IX and Henry III, the most elaborate architectural work in connexion with the dial was put in hand and finally completed under the direction of the eminent sculptor, Germain Pilon, in 1585, who himself contributed terra-cotta statues of Force (or Piety) and Justice to its embellishment.[1]

The historian Rabel refers to this work in the following terms :

> L'an 1585, sur la fin de novembre, fut achevée l'ouvrage du cadran du Palais, lequel avec sa décoration est estimé le plus beau de toute la France. Le conducteur d'icette ouvrage fut Germain Pilon, maître statuaire, et l'un des premiers dans son art, lequel a rendu des ouvrages cy parfaits en notre ville de Paris et autres lieux de France que la mémoire en sera perpétuelle.

Then follows a description of the design and the inscription, composed by Jean Passerat, Professor Royal of Elocution. That which appears below the dial reads :

MACHINA QVAE BIS SEX
TAM JVSTE DIVIDIT HORAS
JVSTITIAM SERVARE MONET
LEGESQVE TVERI

(A mechanism which so justly divides the twice-six hours admonishes us to uphold justice and to observe the laws.)

[1] A reproduction of an engraving in " Merveilles de l'horlogerie " will be found in Britten's " Old Clocks and Watches."

As this motto shows, the dial consisted of an hour circle of twelve hours, as is in use to-day ; whereas de Vick's clock indicated twenty-four hours in accordance with the Italian system.

Here, then, we have the proof that the motion work had been adapted to comply with the latest fashion, together with an hour hand revolving round the circle. But it can hardly be supposed that these were the only changes, or that so elaborate a scheme of external decoration would have been applied to an antiquated and worn-out mechanism.

About a hundred years later, before Julien Le Roy was born, a further restoration of the dial took place under Louis XIV ; and finally the works described by Le Roy were entirely reconstructed by Lepaute in the eighteenth century, when fresh decorated work was substituted for that of 1585, and Pilon's statues were replaced by allegorical figures executed by the sculptor Toussaint.[1]

What then was the mechanism constructed by de Vick? Fortunately, we have in the Science Museum at South Kensington a clock which was discovered in Dover Castle, which most probably resembles it in general construction (see Chapter V).

This view is confirmed by the description of a clock given by the great chronicler Jean Froissart, inspired doubtless by the installation of De Vick's clock in the Palais.

Amongst the poems of Froissart, which have so far been printed, is one entitled " Li Orloge Amoureus " ; and even that was not given *in extenso* until it was reproduced from Scheler's text in vol. i of the Supplement to Kervyn van Lettenhove's edition of Froissart's works, published in Brussels in 1870. The poem, which consists of 1,174 lines, symbolizes, under the guise of a clockmaker and a clock, an ardent lover's adoration of his lady-love. It is written in the

[1] An entirely new clock, with external decoration, was substituted under Napoleon III.

romantic style, so dear to the heart of poets of that age, full of imagery and allegorical hyperbole, and is, be it said, extremely tedious. But, from the horological point of view, it is of surpassing interest, inasmuch as it gives us in the few passages, where the historian abandons the fantasy of the poet, a marvellously accurate picture of the clockwork of the time. Froissart was thirty-two years of age when de Vick had finished his task. It is easy to imagine what intense interest the inauguration of the first public clock in Paris must have created. It was an historical event which could not fail to command the attention of the chronicler. The technical details which the poem contains are so precise and so free from poetical embellishment that Froissart must have derived them from a close examination of the mechanism and probably from de Vick's personal explanation of its action.

The most important passage in the poem has been quoted in several horological treatises ; but in many respects they have not, in my judgment, correctly interpreted Froissart's description. This has been due, as I believe, to the false impression created by Le Roy's account of a clock of much later and more elaborate design. If we read the poem in the light of the Dover clock we shall, I think, come to the conclusion that, as a layman's description of what must to him have been a strange and complicated mechanism, Froissart represents very faithfully the actual clockwork.

I here reproduce, not only the salient extracts of dry description, but everything in the poem that bears direct relation to the clock and its operation, including, therefore, certain imaginative passages which were reproduced for the first time in Scheler's text. I have ventured to attempt a translation of the verses, line by line, hoping that the stigma of *traditore* will not, in this instance, attach to the *traduttore*.[1]

[1] Some obscurities in the text may be attributable to misunderstanding on the part of the original copyist of the mechanical details, resulting in inaccurate transcription of Froissart's manuscript. The accents would be a modern addition.

CI S'ENSIEUT UN DITTIÉ D'AMOUR, QUI S'APPELLE LI ORLOGE AMOUREUS.

1 Je me puis bien comparer à l'orloge,
Car quant amours, qui en mon coer se loge
M'i fait penser et mettre y mon étude,
J'i aperçoi une similitude
Dont moult me doi resjouïr et parer ;
Car l'orloge est, au vrai considérer,
Un instrument très bel et très notable,
Et c'est aussy plaisant et pourfitable ;
Car nuit et jour les heures nous aprent
Par la soubtilleté qu'elle comprent,
En l'absense meïsme dou soleil :
Dont on doit mieuls prisier son appareil,
Ce que les aultres instrumens ne font pas,
Tant soïent fait par art et par compas :
Dont celi tienc pour vaillant et pour sage,
Qui en trouva premièrement l'usage,
Quant par son sens il commença et fit
18 Chose si noble et de si grant proufit.

. . . . .

99 Or vœil parler de l'estat de l'orloge ;
La premerainne roe qui y loge
Celle est la mère et li commencemens
Qui fait mouvoir les aultres mouvemens
Dont l'orloge a ordenance et manière.

. . . . .

113 Le plonk trop bien à la Beauté s'acorde ;
Plaisance rest moustrée par la corde,
Si proprement c'on ne poroit mieulz dire ;
Car tout ensi que le contrepois tire
La corde à lui, et la corde tirée,
Quand la corde est bien à droit atirée,
Retire à lui et le fait esmouvoir,
Qui autrement ne se poroit mouvoir,
Ensi Beauté tire à soi et esveille
La Plaisance dou coer, *etc.*

. . . . .

189 Car Plaisance trop bien à lui s'accorde,
Qui remoustrée est par la propre corde
Que le plonk tire, et dont il fait mouvoir
La mère roe. Ensi m'est-il pour voir.

. . . . .

TRANSLATION

1 I well myself may liken to the Clock,
For when that Love, that dwells within my heart,
Directs my mind thereto and bids me think,
I see therein is a similitude,
Wherefor I must rejoice and pride myself;
For the clock, if well we ponder, is
An instrument most fine and notable,
And pleasant is, besides, and serviceable;
For night and day the hours it lets us know,
Through the subtlety that it possesses,
Even in the absence of the sun:
Whence all the more its mechanism should be prized,
As that which other instruments do not do,
However much by art and compass they be made:
Wherefor I hold him for a mighty man and wise
Who first of all the use of it discovered,
When of his intellect he took in hand and made
18 A thing so noble and of so great profit.

. . . . .

99 Now will I speak of the fashion of the clock;
The first wheel which therein is housed
It is the mother and commencement
Which makes the other movements to turn
Of which the clock has rule and governance.

. . . . .

113 The lead with Beauty doth full well accord;
Plesaunce by the cord is typified
So truly that it might not better be expressed;
For just as the weight draws
The cord to itself, and the taut cord
(When the cord is right well strained)
Pulls it towards itself and makes it move,
Which otherwise could never move,
Thus Beauty draws towards itself, and wakes
The Plesaunce of the heart, *etc.*

. . . . .

189 For Plesaunce doth full well therewith accord,
As is betokened by the cord itself,
Which the lead draws, and so it makes to move
The mother wheel. Thus 'tis plain to see.

. . . . .

201 Et pour ce que ceste roe première
A de mouvoir ordenance et manière,
Par la vertu dou pois que le plonc donne,
Dont, selonc ce, elle dou tout s'ordonne ;
Le plonc le tire, et elle à li s'avance.
Et pour ce qu'elle iroit sans ordenance,
Et trop hastievement, et sans mesure,
S'elle n'avoit qui de sa desmesure
Le destournast et le ramesurast,
Et de son droit rieule le droiturast,—
Pour ce y fu, par droit art ordonnée,
Une roe seconde et adjoustée,
Qui le retarde et qui le fait mouvoir
Par ordenance et par mesure, voir
Par la vertu dou foliot aussi,
Qui continuelment le moet ensi,
Un heurt[1] à destre et puis l'autre à senestre,
Ne il ne doit ne poet à repos estre ;
Car par li est ceste roe gardée
220 Et par vraie mesure retardée.

. . . . .

347 Après affiert à parler dou dyal ;
Et ce dyal est la roe journal
Qui, en un jour naturel seulement
Se moet et fait un tour précisément
Ensi que le soleil fait un seul tour
Entour la terre en un naturel jour ;
En ce dyal, dont grans est li mérites,
Sont les heures vingt et quatre descrites.
Pour ce porte il vint et quatre brochetes,
Qui font sonner les petites clochetes,
Car elles font la destente destendre,
Qui la roe chantore fait estendre
Et li mouvoir très ordonnéement
Pour les heures moustrer plus clerement.
Et cils dyauls aussi se tourne et roe,
Par le vertu de celle mère roe
Dont je vous ai la propriété dit,
A l'aïde d'un fuiselet petit
Qui vient de l'un à l'autre sans moyen ;
366 Ensi se moet rieuléement et bien.

. . . . .

[1] The copyist, ignorant, doubtless, of the action of the foliot, vibrating to and fro, has mistaken " une heure " for " un heurt," which I have substituted for the former word in the text.

201 And by reason that this first wheel
Has in its motion rule and governance
By virtue of the weight given by the lead,
So, therefore, it (the wheel) controls the whole ;
The lead draws it, and it (the wheel) advances towards it ;
And, because it would go without governance
And all too swiftly and without restraint
Had it not something that from its unruliness
Should hinder it and bring it back to rule,
And by its due control should regulate it,—
For this, by proper art arranged,
There was a second wheel adjusted,
Which slows it down and makes it move
By governance and with restraint, to wit
By virtue of the foliot as well,
The which without cessation thus is moved :
One stroke to the right and then one to the left,
Nor may it nor it cannot be at rest ;
Because by it this wheel is checked
220 And by true moderation stayed.

. . . .

347 Next it behoves us of the " dial " to speak ;
And this dial is the journal wheel
Which only in one natural day
Is moved and makes one turn exactly,
Like as the sun doth make a single turn
Around the earth in just one natural day ;
Upon this " dial," whereof the merit's great,
The four and twenty hours are inscribed.
Wherefor it carries four and twenty pins
Which cause the little bells to sound ;
Because they cause the detent to unlock
Which liberates the striking wheel
And makes it move full orderly
So as to show the hours more clearly.
And this " dial " revolves and turns as well
By virtue of that mother wheel,
Whose property I have already told,
By help of a little spindle (? pinion)
Which, naught intervening, goes from the one to the other ;
366 Thus is the motion regular and true.

. . . . .

385 Et ce dyal, qui Doulc Penser figure
Se moet par l'ordenance et la mesure
Que la mère roe d'amours li donne ;
388 C'est à dire, qui bien à droit l'ordonne.

. . . . .

587 Car Doulc Penser, qui continuelment
Me moet le coer, me donne finalment
Par le confort de bonne pourveance,
En tout mon fait matere d'espérance.
Tout ensi que le dyal a manière
De li tourner par la roe première,
Car dou droit tour naturel qu'elle tourne
La roe de Desir à ce l'atourne,
A l'aïde d'un petit fuiselet
Qui nullement ne li fault ne le let,—
Tout ensi Pourveance sans moyen
Ne me poroit fallir pour nulle rien.
Après affiert dire quel chose il loge
En la tierce partie de l'orloge ;
C'est le derrain mouvement qui ordonne
La sonnerie, ensi qu'elle se sonne.
Or fault savoir comment elle se fait.
Par deus roes ceste oevre se parfait ;
Si porte o li ceste première roe
Un contrepois par quoi elle se roe
Et qui le fait mouvoir, selon m'entente,
Lors que levée est à point la destente ;
Et la seconde est la roe chantore.
Ceste a une ordenance très notore
Que d'atouchier les clochetes petites
Dont nuit et jour les heures dessus dittes
Sont sonnées, soit estés, soit yvers,
614 Ensi qu'il apertient, par chans divers.

. . . . .

927 Et pour ce que li orloge ne poet
Aler de soi, ne noient ne se moet,
Se il n'a qui le garde et qui en songne ;
Pour ce il fault à sa propre besongne
Un orlogier avoir qui tart et tempre
Diligamment l'aministre et attempre,
Les plons relieve et met à leur devoir ;
Ensi les fait rieuléement mouvoir,
Et les roes amodere et ordonne,
Et de sonner ordenance lor donne.

385 And this dial, which represents Sweet-Thought,
Is moved by governance and moderation
Which the mother wheel of Love imparts,
388 That is to say, which orders it right well.

. . . . .

587 Because Sweet-Thought, which ceaselessly
Doth stir my heart, gives me at last
By the comfort of good prudence
In all my deeds matter for hope ;
Just as the dial has its way
Of being turned by the first wheel ;
Because, for the proper natural turn it makes,
It is the wheel of Desire causes it to revolve
By help of a little spindle
Which nowise fails nor lets it stay,—
Just so, naught intervening, Prudence
Could not for any reason fail me.
Next it behoves to tell what there is housed
In the third portion of the clock ;
It is the last movement which controls
The striking, so that it may sound.
Now how 'tis made we needs must know ;
By two wheels is this work performed ;
This first wheel carries with it
A weight by means of which it turns
And causes it to move, as so meseems,
When at the appointed time the detent's raised.
And the second is the striking wheel.
To this belongs a very special task,
Namely, to strike upon the little bells
Whereby the aforesaid hours by day and night
Are sounded, be it summer or winter,
614 As appertains thereto, with different tones.

. . . . .

927 And by reason that the clock cannot
Go of itself, nor in any way could move,
Were there no one to keep it and to tend,
Therefore it needs for its especial care
To have a clockman, who both late and early
Shall diligently attend and regulate it,
Draw up the weights and set them to their task ;
'Tis thus he makes them orderly to move,
And both controls and regulates the wheels,
And for the striking gives them governance.

Encores met li orlogiers à point
Le foliot, que ne se cesse point,
Le fuiselet et toutes les brochetes,
Et la roe qui toutes les clochetes,
Dont les heures, qui ens ou dyal sont,
De sonner très certainne ordenance ont,
Mès que levée à point soit la destente.
Encore poet moult bien, selonc m'entente,
Li orlogiers, quant il en a loisir,
Toutes les fois qu'il li vient à plaisir,
Faire sonner les clochettes petites
948 Sans disrieuler les heures dessus dites.

It is not to be expected that, in an allegorical poem, Froissart would have minutely described all the technical details of de Vick's or some other contemporary clock; but we are able in broad outline to gather what were the principal features of its construction.

The going train, as in the Dover clock, and as in the Nuremberg clock (Fig. 7), consisted only of a great wheel and a 'scape-wheel, the latter controlled by a bar-balance or foliot. The same simple construction of a turret clock is illustrated in Giuseppe da Capriglia's "Misura del Tempo,"[1] published at Padua, 1665, as still in use (Fig. 9). Whether the rope passed over a spiked pulley attached to the great wheel, or was coiled round a drum, was a detail to which he would not attach importance. He was only concerned

[1] Capriglia's work, written in 1663, is the earliest treatise on clock-making, with numerous illustrations. As it is a book of extreme rarity it may be of interest to quote the title: "MISURA DEL TEMPO cioè Trattato d'Horologij da Ruota di tre Ordini, Rusticho da Campanile, Pulito da Camera, e Lustro da Petto. Doue con Regole Facili, è Certe s'insegna à Fabricarli di Novo, Et Intender i Fabricati. Materia non più Trattata, ò posta in luce d'alcuno. Opera Curiosa, & utile non solo à chi desidera imparar tal Arte, mà anco à chi si diletta di tenere, è maneggiar Horologij d'ogni sorte, Con la quale ogn'vno può imparar ad accommodarli, & tenerli aggiustati, & in ordine.

"Del P. F. GIVSEPPE DA CAPRIGLIA CAPVCCINO
Peritissimo in tal Arte.
IN PADOVA, MDCLXV."

It is remarkable that, though permission to print the work was granted by the Minister-General of the Capuchin Order on June 7, 1663, Capriglia makes no allusion to the recent application of the pendulum to clockwork by Huygens.

Besides, the clockman accurately sets
The foliot, so that it may never stop,
The spindle too and likewise all the pins,
The wheel, moreover, so that all the bells
Whose hours, which are set upon the dial,
May right truly be arranged to sound,
When at the appointed time the detent's raised.
Quite well, moreover, as I understand,
The clockman, when he has leisure therefor,
And whenever it so pleases him,
Can cause the little bells to strike
948 Without disordering the aforesaid hours.

---

to explain that the weight was the motive power, and that the force transmitted through the cord, to which it was attached, caused the great wheel to revolve. The motion of the latter was then transferred to the 'scapc-wheel. He neglects to mention the intervening pinion, just as he takes no account of the arbors of the wheels and the framework in which they were pivoted. He is only interested in the revolution of the wheels, and in pointing out that the second ('scape) wheel would turn far too rapidly unless some means were applied to check its course. The foliot swinging to and fro and regulating the action is accurately described. The weights on the balance are not mentioned, but he clearly had them in mind, when he tells us that it was part of the clockman's duty to adjust the foliot exactly.

I put forward in Chapter I the suggestion that in the earliest clocks the pallets swung between pins instead of the teeth of the crown-wheel. I have translated Froissart's description as follows :

Besides, the clockman accurately sets
The foliot, so that it may never stop,
The spindle too and likewise all the pins.

The language is obscure ; but, if we are to understand by spindle (*fuiselet*) the verge, then the pins would be those of the 'scape-wheel. Following upon the " setting of the foliot " this interpretation seems to be admissible ; otherwise the controlling action of the verge and pallets, of which the foliot forms an integral part, is left unnoticed. As against this, the allusion may be to the fuiselet which drives the dial-wheel which " carries four and twenty pins." And in the

Figura Nona.
Di tutta la Macchina dell'Orologio da Campanile in profilo, & suoi moti.

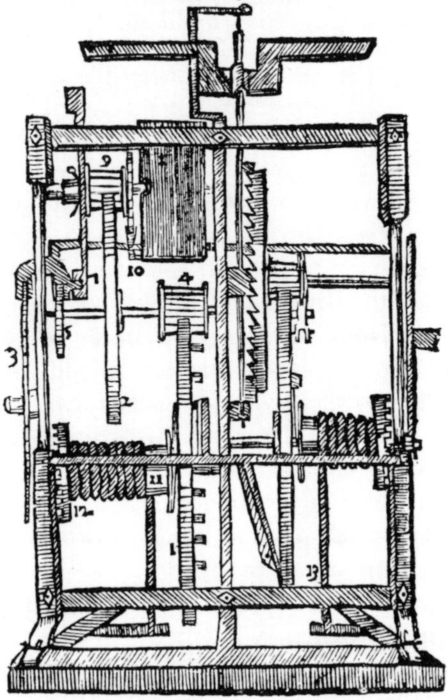

Fig. 9.—Capriglia's Turret Clock.

following line the wheel which regulates the striking may be taken to be either the dial-wheel or the locking plate.

He next proceeds to tell us how the hours were marked ; and here his description has been much misunderstood. These primitive clocks had neither dial and hour circle nor hand. The " dyal," as Froissart correctly describes it, was a wheel driven directly (" sans moyen ") by the great wheel of the going train. I have already described in Chapter I how the end of the arbor of the latter was prolonged through the pivot hole in the frame, and was filed out to form two, three, or more pins. These pins answered the purpose of the pinion or small wheel, which in later days was made fast to the end of the arbor. They engaged in the teeth of the motion or dial-wheel, causing it to revolve once in a given time.

In Froissart's time the day commenced at dusk, and was divided into twenty-four hours. Consequently the " dyal," or *journal*-wheel, had to make one revolution "just as the sun makes a single turn around the earth in one natural day." On the dial-wheel were painted the numerals from I up to XXIV ; and, as it revolved, each figure in turn passed a fixed index. This method of indicating the hour is well shown in the Frontispiece.

The one detail in this part of Froissart's description which is vague is the little " fuiselet," by means of which he tells us that the " dyal " was caused to revolve, connected with the " mother-wheel." I have translated the word " spindle "; but here either his observation or his memory was at fault ; for it is clear that a single pin projecting from the arbor of the great wheel could not have driven the twenty-four-hour wheel. It is probable that the arbor terminated in three pins ; and that the motion-wheel had seventy-two teeth. A smaller wheel would not have carried conveniently the twenty-four " brochetes " or pins, of which he next speaks.

These twenty-four pins would be fixed to the back of the motion-wheel at right angles to it, for the purpose, as he explains, of raising the detent to unlock the striking train at each hour.

This method of dividing the day into twenty-four hours, commencing at dusk, was maintained in Italy, as already mentioned, until comparatively recent times. To alter the clock to suit the change of season formed part of the duties of the attendant, of which Froissart speaks.

It is impossible to say what were the numbers of the two wheels in de Vick's clock; but there were probably ninety-six teeth in the great wheel, as in the Nuremberg and Dover clocks, and the driving weight and train would be calculated so as to make it turn once in an hour. Then with a pinion of three (or three projecting pins) on the end of the great wheel arbor and a "journal" wheel of seventy-two teeth, the latter would revolve once in twenty-four hours.

It is when we come to the striking train that much is left to the imagination. Froissart was able to describe the simple mechanism of the going train and the motion-wheel; but the details of the striking action were either too complicated for a layman, or he did not need them for the purpose of his poem. He tells of the great wheel with its cord and weight, and of a second wheel—the "roe chantore"—which governed the proper sounding of the hours; but he is silent as to any other wheel or fly for controlling the speed of the strokes; nor does he mention the hammer. Clearly he understood that the pins on the dial-wheel served to raise the detent, and thus to bring the striking into action; but it is not clear to me whether the "roe chantore" is intended to indicate the "second wheel," with pins in its periphery to raise the tail of the hammer, or the "locking plate." I incline to think that the latter is meant, because he says that by lifting the detent the clockman could at pleasure cause the bells to strike. He adds that this could be done without disarranging the hours. That is true, so far as the position of the dial-wheel is concerned; but I think that he must have had in mind the case of the clock having failed to strike, as may happen when the locking plate and detent are badly made and adjusted. The clockman would then raise the detent, and allow the clock to strike until the proper notch in the locking plate was reached.

So far as is known, de Vick's clock struck on one bell; whereas Froissart speaks of striking on little bells. Possibly the poem was written before de Vick had received the

King's definite instructions as to the method of sounding the hours, and before it had been decided that there should be a single large bell, to be cast by Jouvence. If this is the explanation, it may also account for the meagre description of the striking train ; or the detail may have been filled in from Froissart's recollection of some other clock which had a ting-tang action. The second part of the Chronicles deals principally with events in Flanders, where many towns already possessed clocks which may have had a chime on "little bells" before the hour struck.[1]

I owe to my friend, General J.-B. Dumas, the preservation of a remarkable inscription which existed in 1731, possibly the date of Le Roy's examination of the clock. The legend, in six lines in Gothic character, not only recorded the fact that Charles V had installed the bell, but in the form of a cryptogram it gives the exact date. As originally written, certain letters were of larger size than the rest, these constituting the chronogram. The hidden meaning which this inequality conveyed was not appreciated, when in course of time the verses came to be repainted ; and the uneven lettering was doubtless attributed to want of skill on the part of the original scribe, whose clumsiness must not be perpetuated.

The lines as originally written read as follows :

CHARLES ROI VOLT EN CE CLOCHER
CETTE NOBLE CLOCHE ACROCHER
FAITE POVR SONNER CHACVNE HEVRE
LA DATE ESDITS TROIS VERS D'ASSEVR
PAR JEAN JOVVENCE FVT MONTEE
QUI DE CET ART EST RENOMME.

[1] Froissart was present in Milan in 1372 for the wedding of the Duke of Clarence to the daughter of Galeazzo Visconti, when he cannot have failed to see the wonderful clock of 1335, I have which described in Chapter II.

The capital letters in the first three lines give the date, taking C as 100, L as 50, V as 5 and I as 1, thus :

11	C's	=	1100
5	L's	=	250
4	V's	=	20
1	I	=	1
			1371

Poor de Vick ! His celebrated masterpiece failed to give the results that his patron hoped for, as an accurate time-keeper and as a trustworthy herald of the passing hour. Its pristine glory departed, until at last it became an object of derision, popularly expressed in the ribald rhyme :

> C'est l'horloge du Palais ;
> Elle va comme ça lui plaît.

And I think that in this little skit we have conclusive proof that the clock described by Le Roy was not that which, as a result of eight years' labour, was completed in 1370 by Henri de Vic. As a parallel case one may quote the Hampton Court Clock, which, after similar vicissitudes, received a new movement in 1879 ; and the same may be said of the celebrated Strasburg clock.

The subsequent activities of de Vick are unknown ; but the clock erected at Metz in 1391 has been attributed to him. One thing, however, is certain, that he did not retain the favour of Charles V, who would appear to have taken much interest in clocks. He was succeeded by Pierre de Sainte Béate, to whom reference has been made in Chapter III.

DOVER CASTLE CLOCK

By courtesy of the Science Museum

CHAPTER V

THE DOVER CASTLE CLOCK

As I have referred to this clock in order to illustrate Froissart's poem and the mechanism of de Vick's clock, I think that it will be of interest to state what little is known about it.

The first mention of it appears in Admiral W. H. Smyth's article upon an astronomical clock, made in 1525 by Jacob Zech of Prague, in the possession of the Society of Antiquaries, contained in vol. 33 of *Archæologia*. Speaking of clocks governed by a foliot balance the Admiral says : " Mr. Rowell of Oxford assured me that he had seen one among some rubbish in a turret in Dover Castle." As a result Admiral Smyth, accompanied by Mr. B. L. Vulliamy, paid a visit to Dover, when, with the permission of the authorities, the clock was removed from its hiding-place, set up and cleaned of the dirt of ages and rubbed over with linseed oil. He says : " We were delighted to find, nearly entire, an unsophisticated old clock bearing the date 1348 and the mark  ; the wheels, fly vanes and frame being of wrought iron, and it was wound up by means of small spokes on the end verge of a cylindrical wood barrel. Mr. Vulliamy took the necessary measurements for a professional description of this remarkable relic, with a view to its being published."

I have searched in vain for any such record.

The clock is still in going order, and is one of the most interesting exhibits in the horological section of the Science Museum at South Kensington. I have counted the trains, and they are as follows :

Going train	*Striking train*
Great wheel 96, pinion 7	Great wheel 72, pinion 9
Crown-wheel 33	Second wheel 72, pinion 8
Foliot 28″ long	Fly

The wooden barrel, which carries the cord, is loose on the arbor of the great wheel, to which it is coupled by means of a spring click, catching against a spoke of the wheel, so that the arbor can only turn in one direction. The cord is wound on the barrel by four spokes fixed to it inside the frame. The arbor is prolonged through the front of the frame, and terminates in a squared end, to which, no doubt, was fixed a pinion for driving the motion-wheel, for which there is a pivot hole higher up. A square hole lower down in the frame probably served to secure a bridge for carrying the front pivot of the motion-wheel.[1]

The hammer work is missing, as also the dial. Mr. C. Octavius S. Morgan, F.R.S., in "Notes on the Clock,"[2] mentions a hand 19 inches long, but I have never seen it.

Mr. Morgan considers that it was originally set up in the old church, and that, when the church fell into ruin, it was brought to the castle and never re-erected. Its wonderful preservation would be due to its having been stored away as lumber until its discovery by Mr. Rowell.

There has been much controversy as to the age of the clock. It was again examined in 1851 by the late Mr. Albert Way and Mr. Franks, both of whom, from the absence of any Gothic ornamentation to the frame, were of opinion that it was not of earlier date than 1450 to 1500, and possibly later. Mr. Morgan agreed with this view, and also pointed to the absence of "lantern" pinions as negativing very great antiquity.

Bilfinger[3] discusses the question from a different point of view. He considers that the "modern hours" were not adopted in England before the latter part of the fourteenth century, and he refers to the striking-clock made for Edward III (see p. 35) which the Monk of Malmesbury described as the first to distinguish the twenty-four hours. Now, the locking plate of the Dover clock is cut for twelve

[1] The suspension cord of the verge is not shown in the illustration.

[2] *Horological Journal*, January 1917.

[3] *Op. cit.*, pp. 219–20.

hours, and we have no certain evidence of the 2×12 hours system in England before the end of the fourteenth century (see p. 38).

It has been stated, but on what authority I do not know, that the clock was made in Switzerland. As to this, Bilfinger remarks that there is no evidence of any striking-clock in that country until twenty to thirty years later than 1348. He suggests that the clock may have come from Italy ; but that seems unlikely, seeing that the twenty-four-hour system was general there.

As regards the date on the clock, upon which the whole controversy rests, no one who has examined it since its discovery has been able to find any trace of it. It is, however, unlikely that Admiral Smyth, the Secretary of the Society of Antiquaries and a distinguished archæologist, and Mr. Vulliamy, the eminent clockmaker, could have been mistaken as to the presence of a date when they first examined it. The probability is that as a result of subsequent cleaning the outer skin of the rusted iron peeled off, and that with it the date disappeared. Or can it be that they misread the faint inscription for 1548 ? In any case, we should not expect to find a date in Arabic numerals in the middle of the fourteenth century.

Reviewing the question from every point of view, the balance of probability seems in favour of assigning a date much later than 1348, and of regarding the clock as one which might well have been constructed in the sixteenth or even in the seventeenth century. Capriglia's design (Fig. 9) shows that such a simple two-wheel going train was still in use even in 1663.

Recent research at the Record Office by Lieut.-Commander Rupert T. Gould,[1] ranging from Edward III to Henry V, has resulted in the discovery of only one reference to a clock in Dover Castle. From this it appears that for

[1] Author of "The Marine Chronometer ; its History and Development," London, 1923.

eleven years, from some date between 1404 and 1407, a custodian of a clock was paid 13*s.* 4*d.* per annum for his services. This proves conclusively that there was a clock in the castle or in the church in the early part of the fifteenth century; but we cannot therefore assume that it was the clock now extant in respect of which the payments were made.

Since the above notes were written the discovery has been made of a remarkable clock, included amongst the lumber at the sale at Cassiobury Park, the seat of the Earl of Essex in Hertfordshire. By good fortune it has passed into the possession of Mr. Percy Webster, than whom no one is more competent to appreciate its historic value. The clock is a somewhat smaller counterpart of the Dover clock, the frame measuring 22 × 17 × 22 inches. It is complete with the motion-wheel, the foliot balance and the hammer, whereas the Dover clock has several parts missing. The trains have the same numbers as the Dover clock, except that the crown-wheel pinion is 6 and the balance is 19 inches long. After a thorough cleaning of the dust and rust of ages it was found to show but little sign of wear, and it is now erected and in going order at Mr. Webster's residence, not far from its old home.

Evelyn tells us (April 18, 1680) that he paid a visit to "Cashioberie" on the invitation of the Earl of Essex to inspect the new house which was being completed on the site of the old mansion. Everything was being executed in the latest style. We may therefore fairly assume that the clock had been in use in the old buildings, and was now considered to be too antiquated for what Evelyn calls "a very noble palace." This assumption is all the more justified, because we know that a turret clock, fitted with the new anchor escapement, had been made as early as 1671 (see Chapter VIII). The old clock would then, like the Dover clock, be relegated to some lumber room, where it was to repose for 250 years.

From its excellent state of preservation we may conclude that at that time it was not of great antiquity, and it may even date from the seventeenth century. Mr. Webster would put the date as certainly prior to 1625. However this may be, I am of the opinion that the Cassiobury clock confirms my suggestion that the Dover clock is of a much later date than that which has been assigned to it (1348).

A more recent discovery has been that of a clock of similar construction in a farm building at Quickswood (Herts). It has the same smith's punch mark as the Cassiobury clock, but unfortunately in this case the foliot balance has been replaced by an anchor escapement. The Dover clock and Mr. Webster's are, so far as is known, the only clocks existing in England in their primitive condition. All the celebrated clocks at Wells, Exeter, Wimborne, Ottery St. Mary, etc., have been modernized. Need we wonder? The crude blacksmith's work of those early days, with the coarsely cut wheels and pinions, was not conducive to lasting wear, as well exemplified in the fate of de Vick's clock.

GALILEO

From the painting by Sustermans. By courtesy of the Bodleian Library, Oxford

CHAPTER VI

THE ROMANCE OF THE PENDULUM

ACCURATE time measurement was first made possible by the adaptation of a swinging pendulum to clockwork ; and, as the solution of this problem may rightly be regarded as one of the great achievements of science, it is therefore a matter of no small interest to determine to whom this notable invention should be attributed. " *Palmam qui meruit ferat.*"

The timepiece, as we know it, is the result of three principal discoveries. The first step was the invention of toothed wheels for the transmission of power, which, rightly or wrongly, has been attributed to Archimedes in the third century before our era. Contrate wheels and lantern pinions were also devised at an early date. Claude Perrault, in his edition of the Works of Vitruvius,[1] gives a copper-plate illustrating the description of the elaborate clepsydra of Ctesibius (*ca.* 135 B.C.). The motive power of water was regularly employed thereafter for driving trains of wheels, giving the motions of the heavenly bodies. The ground was thus already prepared for the construction of simple and elaborate mechanisms when a way should be discovered of controlling the force obtained from a falling weight or a coiled spring.

The invention of the verge escapement surmounted by a cross balance, regulated by small weights or riders, was the next great step in time measurement, as already described.

For three centuries and more the imperfections of this form of escapement were perpetuated. Meantime, the great development of astronomical science, due to Copernicus, Tycho Brahe and Kepler, called for a better means of measuring time, whilst the immense expansion of terrestrial discovery by Prince Henry the Navigator, Vasco da Gama, Cabot, Columbus and others in Asia, Africa and the " New

[1] Paris, 1684.

World " made it a matter of pressing necessity to devise some method of ascertaining the longitude at sea.

In 1598 the King of Spain had offered a reward of 1,000 crowns for a means of determining longitude, and this was followed by an offer of 10,000 florins by the States-General of the Netherlands.

Such then was the state of affairs when Galileo's experiments with a suspended weight, and its isochronous vibration, as he supposed, whatever the angle of the arc, became known to the world. Already at the age of seventeen he had made his discovery in 1581 from observing the swing of a lamp suspended in the Cathedral of Pisa, and had thus laid the foundation of the third step in the achievement of exact time measurement. It is remarkable that so many years should have elapsed before this momentous discovery was put to practical account.

In 1657 the celebrated mathematician Christiaan Huygens submitted his pendulum clock to the States-General by his assignee. He had assigned his rights in the invention to a skilful clockmaker of The Hague, Salomon (or Samuel) Coster, to whom a patent (*octrooi*) was granted on June 16, 1657, for twenty-one years.[1]

In May 1923, while on a visit to the Rijksmuseum at Amsterdam, I remarked a timepiece in a small black rectangular case, fixed to the wall in an inconspicuous position and unlabelled. The dial, as was usual in early Dutch clocks, was covered with black velvet, and it had a plain hour circle with a single hour hand. The Director most kindly gave permission for it to be unscrewed from its place. The back proved to be a plain piece of deal fastened

[1] In the State Archives at The Hague I have seen the original Minute of the States-General of June 14, 1657, when Coster's " Request " was referred to a Committee to inspect the clock, and a further Minute of June 16, when it was approved. The *octrooi* bears the same date, and a copy of it is entered in the General Register. But the large volume of Requests of 1657 does not, unfortunately, contain Coster's Memorial, nor has the description by Huygens of the clock of that date come to light, on which the " Request " must have been founded.

Fig. 10.—Dial of Bracket Clock by Salomon Coster.

to the frame with antiquated nails. These were with some difficulty extracted, and there was then revealed a plain *spring* movement [1] without any adornment of engraving

[1] Huygens himself writes in "Horologium" that "artificers, once informed as to the principle of my invention, will find the way to apply it to clocks of different types. . . . I have myself seen in the workshop of the man, to whom in the first place I had recourse for manufacture, clocks of this description having as their motor not a weight but a spring."

Fig. 11.—Movement of Coster's Clock.

on the back-plate, but inscribed—

SAMUEL COSTER
—HAGHE
met privilege 1657.

Here then was a clock made in the year of the grant of the patent.[1]

The escapement consists of a *horizontal* crown-wheel and verge, to the end of which is attached a crutch carrying in its fork the pendulum rod. The pendulum itself is suspended by a double thread from a cock or bracket, to which are attached curved cheeks, but these were not yet "cycloidal."

The Rijksmuseum possesses another and more elaborate clock by Salomon Coster

[1] This treasure was not included in the exhibition of clocks and watches held at The Hague in July and August, 1929, in celebration of Huygens' tercentenary, nor any other clock by Coster. Since the reconstitution of the Museum it has been missing.

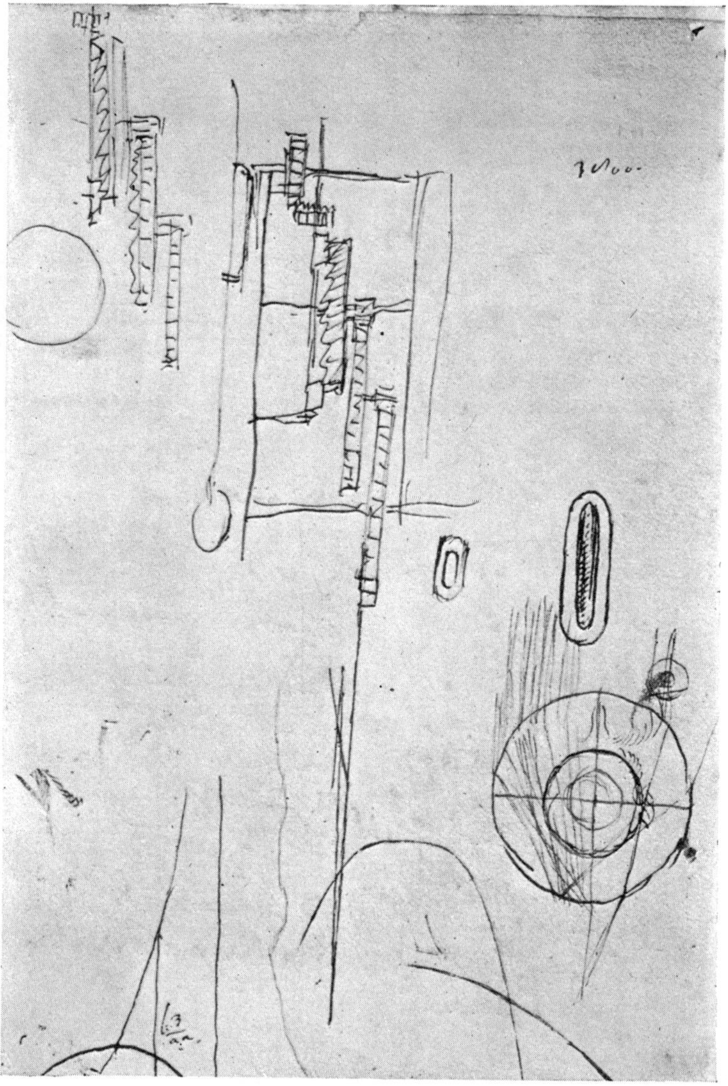

Fig. 12.—HUYGENS' FIRST DRAFT OF A PENDULUM CLOCK

By courtesy of the University of Leiden

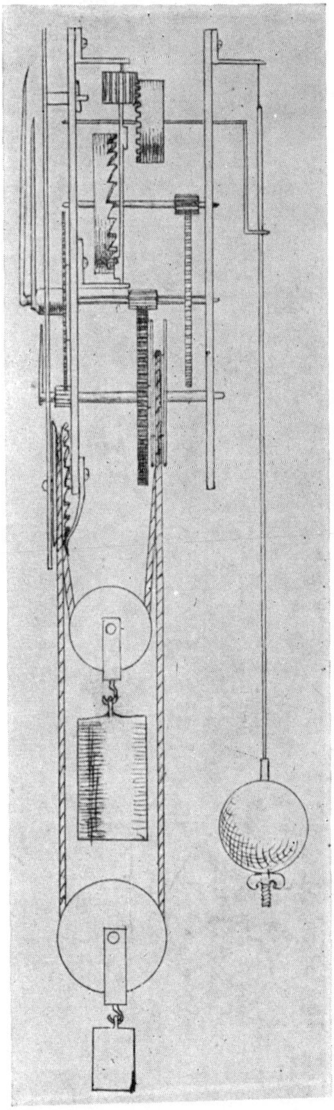

Fig. 13.

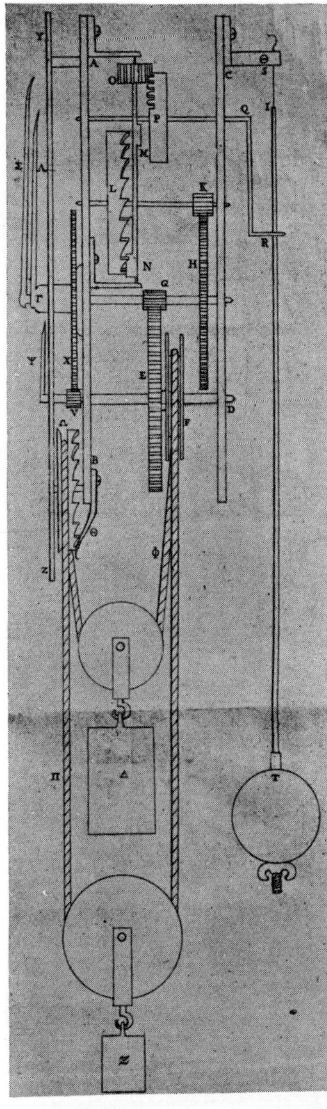

Fig. 14.

with a minute hand, but otherwise made to the same design, as shown in Figs. 10 and 11. The hour circle is mounted on black velvet, and the case is ebonized.[1]

The smaller clock was probably one of the first made after the grant of the patent; for Coster seems finally to have discontinued the use of Samuel as his first name.

Huygens, in 1658, in a small Latin treatise entitled "Horologium,"[2] described the mechanism of his pendulum clock, accompanied by an illustration. This shows that he had already modified his original design. A rough sketch, showing his first conception, made in February 1658, occurs in his Manuscript Book "A," of which I am permitted to give an illustration (Fig. 12).

This was then developed into the finished drawing in his Note-book (Fig. 13), which is placed side by side for comparison with the design (Fig. 14) prepared for the engraving in "Horologium" (but without the shading), where it measures 10 inches over all.

It will be observed that he has introduced a *vertical* crown-wheel, with a pinion O at the top of the verge, which engages in a contrate wheel P, only toothed for half its circumference, this latter wheel carrying the crutch.

Petit,[3] in a letter dated October 18, 1658, rightly suggested that these two wheels might be suppressed, to which Huygens replied that he had introduced them to reduce the arc of vibration of the pendulum, which consequently was less subject to irregularity. He says:

> If with the crown-wheel alone short vibrations were possible, the wheel and pinion might be dispensed with, and the crown-

[1] The clock is now on loan to the Nederlandsch historisch natuurwetenschappelijk Museum at Leiden, where many interesting relics of Huygens are on exhibition.

[2] A French translation, by M. Léopold Reverchon, was published in "*L'Horloger*" of Paris for September 1922. Already in 1660 Huygens was working on a revised and enlarged edition, but by 1662 he had abandoned the intention.

[3] Pierre Petit (1594–1677), mathematician, astronomer and physicist, an intimate friend of Pascal.

wheel and pallets could be placed horizontally, as you say. It is thus in fact that the first of all these clocks was made, which I am still using, and numbers of others have been made after that model. But since then I have preferred that of which I have given the illustration, as not requiring the two cheeks.

It was not until December 1659 that Huygens discovered that the pendulum, in giving isochronous vibrations, did not describe the arc of a circle, but followed a cycloidal curve, a fact unknown to Galileo. This fact he demonstrated later on mathematically in his great work "De Horologio Oscillatorio," published in 1673. In order, therefore, to overcome any circular error, he caused the flexible suspension of his pendulum in all his later clocks to swing on either side against brass "cheeks," formed in a cycloidal curve. Theoretically, the idea was ingenious; but in practice it proved not only unsuccessful, but actually detrimental on account of friction and banking.

The further development of Huygens' designs will be described in the chapter on the "Longitude"[1] (p. 143).

As soon as a specimen cf Coster's clock in 1657 and subsequently a copy of "Horologium" reached Florence, it was not surprising that the disciples of Galileo should resent an attempt, as they regarded it, to rob their illustrious master of the great glory of being the first to elaborate the control of clockwork by means of a pendulum. Charges of plagiarism were made, and an acrimonious disputation ensued, which was not confined to the home of Galileo, and has been the subject of argumentation again and again up to the present day. The purpose of this chapter is to relate the story in the light of Huygens' correspondence, the Medicean archives at Florence, of my researches in Holland, and of the perusal of much foreign literature on the subject.

[1] It would in any case be impossible to recapitulate within the compass of this book all the ideas and sketches which are recorded in Huygens' Manuscripts. The interested reader will find them set out at length in the forthcoming vol. xvii of the *Œuvres complètes*, now being edited by Dr. J. A. Vollgraff.

Dr. Johann Becher[1] in his treatise "De nova temporis dimitiendi ratione Theoria," dedicated to the Royal Society in 1680, affirms that Count Lorenzo Magalotti, Resident of the Grand Duke of Tuscany at the Court of the Emperor, told him the whole history of these pendulum clocks, and denied that Huygens was the author of them. Nay more, he declares that one Treffler, clockmaker to the Grand Duke, had told him that "he had made the first pendulum clock at Florence by command of the Grand Duke, from the directions of his mathematician Galileus à Galileo, a pattern of which was brought in to Holland."[2]

Becher also quotes a statement made to him by the mathematician to the Elector of Maintz that he had seen at Prague a pendulum clock made by Justus Borgen, clockmaker to the Emperor Rudolph, which clock Tycho Brahe used for his astronomical observations.[3] If the latter claim were well-founded, it would seem, from Tycho Brahe's statement quoted by Hevelius, that the clock was not a success (see p. 116).[4]

In France the Abbé de Hautefeuille, writing in 1675,[5] in opposition to the grant of a patent to Huygens for the invention of the balance spring of the watch, reproaches Huygens roundly with having taken credit for the discoveries of Galileo and his son Vicenzio with regard to pendulums. He alleges that Vicenzio having communicated his invention to his friends, they divulged it, and that it came to the knowledge of Huygens ; that Huygens, with his penetrating intelligence, perceived that the inventor had not grasped the value of the discovery, and that having perfected it, he thought himself entitled to claim it as his own.

[1] Johann Joachim Becher (1635–82), physician and chemist.

[2] According to the statement in van Swinden's Memoir, quoted later, this clock was still in existence in Pisa in 1774.

[3] "The Artificial Clockmaker," by Wm. Derham, London, 1696.

[4] According to Thomas Reid, Justus Borgen was a Swiss. Britten gives his name as Jobst Bürgi (1552–1632), and states that he was appointed clockmaker to Rudolph II in 1602.

[5] "Factum de M. l'Abbé de Hautefeuille touchant les pendules de poche contra M. Huygens."

Huygens had already replied in 1673 to the accusations that had been levelled against him in his work, "De Horologio Oscillatorio"; but our own Dr. Robert Hooke, in a review of this work, did not hesitate to brand Huygens as a plagiarist in unmeasured terms.[1]

Although the controversy thus raged during Huygens' lifetime, the disputants had nothing better to rely upon than hearsay evidence and the statement, made by the Accademia del Cimento in Florence in 1662, that a pendulum clock of Galileo's design had been accomplished by Vicenzio in 1649.

The controversy was resuscitated in 1817 by Mr. J. H. van Swinden in consequence of the publication in 1811 of a document written by Vicenzio Viviani in 1659, and again on the fresh discovery, by Prof. Albèri of Florence in 1856, of the actual drawing of Galileo's clock, in the remarkable circumstances about to be related.

In view of this discovery Prof. Albèri[2] addressed a request to M. J. B. Biot (1774–1862), the eminent French physicist, to present to the Académie des Sciences in Paris an account which he had written in reliance on the documents that had come to light, to prove that Galileo was the true inventor of the pendulum clock, executed by his son in 1649, and that to him must be ascribed the honour of having enriched astronomy with so useful an invention.

M. Biot, however, utilized the documents to champion the cause of Huygens. This historical dissertation appeared in the *Journal des Savants* in November 1858, and was reproduced in the *Revue chronométrique* in January 1860, with some excisions.

Whatever conclusions may be drawn from the article, the facts related are of great interest; whilst the history of the discovery of Viviani's Report and drawing is in itself a

[1] See later on his unpublished notes for a lecture amongst his MSS. in the British Museum (p. 167).

[2] Editor of Galileo's Works in 16 vols. (1842–56).

romance ; and I therefore give a résumé of M. Biot's contribution, as follows :

Galileo's experiments with suspended weights led him to the conclusion that if two balls of equal weight were suspended on cords of equal length and they were set in motion at the same time, the one with a swing of 80 degrees or more from the vertical and the other with a swing of only 2 or 3 degrees, they would perform simultaneously the same number of vibrations.[1]

Starting from this theorem, Galileo suggested to the States-General of Holland (in March 1636) the construction of a mechanism for ascertaining the longitude at sea. His idea was to attach to a solid pendulum a stylus which, each time when the pendulum passed the vertical, would strike an elastic bristle fixed at one end perpendicularly to the plane of oscillation and free at the other end. A horizontal crown-wheel was to be placed below this bristle close to the pendulum, so that, on the bristle receiving the impact of the pendulum stylus, it would strike the perpendicular side of the tooth, and drive it forwards, but from its elasticity it would then travel up the concave surface of the other side of the tooth, ready to repeat the operation at the next impact. Such was the basis of the counter that he proposed ; but, without stopping to consider the mechanical difficulties of such a delicate device, he went on to suggest the connexion of this crown-wheel to a regular train of clockwork.[2]

Plainly he had not attempted to put his system to the practical test of construction, because he adds :

It would be a waste of time to occupy Your Lordships' attention any longer with the details. You can command artists of the utmost skill in the manufacture of clocks and other excellent mechanisms. They have only to know that

[1] (" Nouveaux dialogues sur la mécanique," Leyden.) This statement was subsequently proved by Huygens to be inaccurate, the respective deviation at 80 degrees and at the minute vibration being in the proportion of 34 to 29.

[2] Hitherto astronomers had been obliged to count the beats of a pendulum kept in motion by the hand.

the pendulum gives vibrations of exactly equal duration, whether the arc be great or small, in order to devise methods of construction of greater precision than any that I could devise.

The letter containing these proposals was written on June 6, 1637,[1] but they were found impracticable ; and, after much negotiation, the scheme was abandoned.[2]

Galileo died on January 8, 1642.

In 1658 Huygens published in Holland his treatise, entitled " Horologium," giving a practical description of his application of the pendulum to clockwork.

The artist whom Huygens entrusted with the execution of his model quickly grasped the idea of utilizing a spring instead of a weight for driving the train.

Huygens in his book gives " the end of 1656 as the date when he first thought out a new method of measuring time ; and a few months later, having divulged it in Holland, it could not be doubted but that it would soon spread far and wide, by reason of its own utility, inasmuch as several examples of the new mechanism had already been manufactured and dispatched in different directions."

Huygens does not fail to render just homage to Galileo's " inventive sagacity," in having taught astronomers the way to measure time by the swing of a pendulum, but the labour of maintaining the swing by hand and of counting the vibrations one by one was exceedingly laborious.

Huygens' invention was the subject of universal

[1] To Admiral Reael, President of the Committee for their examination.

[2] This statement is misleading. Galileo entrusted the negotiations to his French advocate, Elia Diodati, who was in correspondence on the subject with Christiaan's father, Constantijn Huygens, the Secretary to Prince Frederic Henry of Orange, the Stadholder. Nelli, in his " Life of Galileo," quotes a letter from Diodati of May 15, 1637, to the States-General, which was presented to Constantijn Huygens. This resulted in Galileo's proposals of June 6, after which the negotiations dragged. But Galileo did *not* let the matter drop ; for we find Diodati still corresponding with Constantijn in February and April 1640.

From this it is clear that Galileo was still hopeful of success. His mind (in his blindness) was still at work on the problem, which at last took shape in the last year of his life.

Galileo's sole reward was a gold chain !

admiration and gratitude on the part of astronomers. In Italy alone scientific jealousy and national feeling strove to rob the benefactor of his due. It seemed impossible that Galileo, their glorious and almost only representative in astronomical science, should not have thought of so simple, but at the same time so useful, an invention. Old memories were revived to prove that he had had the same idea in the last year of his life, and that he, therefore, and not Huygens, was the real inventor.

In the Minutes of the Accademia del Cimento of August 11, 1662 (but not printed until 1667) there appears an entry, which states, after referring to Galileo's simple pendulum kept in motion by hand, that "for experiments requiring nicer precision it was judged advisable to apply the pendulum to the clock which Galileo, before anyone else, had devised, and his son Vicenzio had put into practice as early as 1649, so that from the force of the weight or the spring the pendulum is always obliged to fall from the same height."

Unfortunately, no mention is made as to the nature of the escapement, and the illustration accompanying the text shows nothing but a closed drum from which depends a pendulum, nothing of the interior being visible.

The publication of these Transactions deeply wounded Huygens, inasmuch as they presented him in the light of a plagiarist. In his great work, "De Horologio Oscillatorio, sive de motu pendulorum ad horologia aptato," published in Paris in 1673, Huygens warmly repudiated the implication. He says that it was sixteen years ago that he had published his Treatise,[1] and that it seems beyond the bounds of credence that so useful an invention could have remained unknown for eight whole years, until he published it himself. If it was kept a secret, no imputation could attach to him, unless it could be proved that he alone was acquainted with the secret.

[1] I.e. "Horologium," 1658.

Huygens sent, through the French Ambassador to the Court of Tuscany, a copy of his work to Leopold dei Medici, brother of the reigning Grand Duke Ferdinand II (1610–70). Leopold had been a disciple and ardent admirer of Galileo, and it was he who with Viviani, Torricelli and other friends of Galileo had founded the Accademia del Cimento in 1657, of which he was the president. On his being made a Cardinal in 1667 the institution came to an end.[1] Leopold had maintained relations in the past with Huygens in regard to the latter's discoveries in astronomy, physics and mathematics. It was therefore natural that Huygens should accompany his presentation with a letter in which he reminded the Prince that he had sent him at the time a copy of his " Horologium " containing an illustration of his clock, before any suggestion that a similar attempt on the part of Galileo had been given to the world.

In his reply the Cardinal said that he believed in all sincerity that it had never come to Huygens' knowledge that " the idea of applying the pendulum to a clock had also occurred to Galileo, as this was known by very few persons, and as Galileo himself had not carried anything of this project to a state of practical accomplishment, as is apparent from the little that was produced and sketched out (*abbozzolato*) at the hands of his son."

Vicenzio Galilei died in 1649. His widow died in January 1669, appointing Viviani as her executor. In the inventory of her effects, which is still preserved, appears the entry : " An iron clock, unfinished, *with pendulum*, the first invention of Galileo." What must

[1] The Accademia del Cimento, a society of men of science, similar to our own Royal Society, was founded at the suggestion of Vicenzio Viviani. The members were Viviani, Alfonso Borelli, Carlo Rinaldini, Allesandro Marsili, Paolo and Candido (brothers) Del Buono, Antonio Oliva, Francesco Redi, Carlo Dati, with Count Lorenzo Magalotti as Secretary. Correspondents at Rome were Falconieri and Ricci.

On May 8, 1661, a letter from Prince Leopold was read at the meeting of the London Society, not yet incorporated, and a Committee was " appointed to correspond constantly with that Prince."

have been the moral bias of the Academicians of the Cimento in venturing to state in their published Transactions in 1667 that the pendulum clock of Galileo had been *put into practice* by his son as early as 1649 ; as being the original or model of that which they had used in experiments which demanded greater precision than a simple pendulum, and going so far as to fortify these assertions by an illustration deemed to represent it, but which only shows the outer case ! We are to suppose that these disciples of Galileo had kept the fact secret, and only after long years sought to repair the wrong that their long indifference had inflicted ! Viviani, living with Galileo at Arcetri, knew everything about the mechanism of the clock. He remained in intimate relations with Vicenzio, who had been charged by his father to execute the work. Yet, at his death in 1649, Vicenzio left it in a state of sketch, and kept the secret buried for seven whole years. And Viviani realized the importance so little, that in his " Life of Galileo," published in 1654, he does not mention the clock or even the idea !

By an extraordinary train of circumstances a letter of Viviani of August 20, 1659, to Prince Leopold subsequently came to light, a letter which Viviani never divulged during his long life. In it he relates his personal acquaintance with Galileo's invention from its first conception up to the time when its execution was lamentably abandoned. The story of the letter is this. Viviani at the time of his death in 1703 at the age of eighty-one possessed all Galileo's unpublished manuscripts and a quantity of his scientific correspondence. The Roman Inquisition had become exceedingly powerful in Tuscany ; and Viviani might reasonably fear that the papers would compromise him, or lead to their destruction.

They and the letter in question were hidden away in his house in Florence at the bottom of a corn-bin (*buca di grano*), and there remained concealed until the death in 1737 of his nephew and heir, who probably knew nothing of their value or possibly of their existence. Having discovered this mass of papers, his heirs sold them by weight to a pork butcher in Florence, who used them for wrapping up his wares. By rare good fortune one of the parcels fell into the hands of Clemente de Nelli, a Florentine Senator and a zealous champion of the literary glory of Italy. He recognized a letter of Galileo ; and, following up the trail, he recovered for a mere song all that remained of the papers. This was in 1739. Nelli at once set to work to include his treasure trove in a work entitled " Vita e Commercio di Galileo." Lack of time prevented its completion until 1790, and it was not until shortly after his death that the work was printed in 1793 in Florence, but disguised under the imprint of Lausanne. Even so the work was not published until 1811, when Viviani's letter was first given to the world ! The letter was not in Viviani's handwriting, but the corrections were written by him.

As the letter refers to an illustration representing a pendulum clock, devised by Galileo, Nelli reproduced a drawing included amongst the papers, which he supposed to be that mentioned in the text. But it proved to be a copy of the clock illustrated in Huygens' " Horologium Oscillatorium " ! [1]

Signor Albèri, on making a final search, discovered the actual drawing, which had been folded up amongst other scraps of paper, and included a reproduction of it in vol. xvi of his edition of the works of Galileo. The

[1] Nelli (p. 688) gives the entire credit to Huygens, as the first to apply the pendulum to clockwork, and reproduces Huygens' design and full description taken from " Horologium " of 1658. The following chapter consists of Viviani's complete letter, and facing it is a reproduction of Huygens' design of 1673. On this he comments as follows : " The reader may compare it (i.e. the first illustration) with the following one, which latter, under the direction of Galileo's son, was imperfectly executed in 1649, and by command of Prince Leopold de' Medici was described by the celebrated geometrician Vicenzio Viviani." He supposes Huygens' perfected mechanism to be Galileo's failure ! He had not discovered Vicenzio Galilei's actual drawing.

letter itself is printed *in extenso* in vol. xiv, p. 339. Viviani refers at length to Galileo's discoveries as to the simple pendulum, his negotiations with Spain and subsequently with Holland for the adoption of his proposals in regard to the measurement of longitudes. He then continues :

One day in the year 1641, whilst I was living with him in his country house at Arcetri, *I recollect* that it came into his mind that the pendulum could be adapted to clocks driven by weight or spring, in the hope that the perfect natural equality of its motion would correct the imperfections of mechanical construction. But, being deprived of sight and unable himself to execute the plans and models which would be required to ascertain which would be best adapted for carrying out this project, he communicated his idea to his son Vicenzio, who had come out one day from Florence to Arcetri. They had several discussions on the subject, with the result that they fixed upon the method, of which the accompanying drawing is a copy[1]; and they decided to proceed at once with its execution, in order to determine what were the difficulties, which, as a rule, in the construction of machines, a theoretical design does not reveal. But Vicenzio, being desirous to construct the instrument with his own hands, for fear lest the artificers who might be employed should divulge it before it had been presented to the Grand Duke and to the States-General of Holland for the measurement of longitudes, kept putting off its execution, and a few months later Galileo, the author of this admirable invention, fell ill, and died on January 8, 1642. As a consequence, Vicenzio's enthusiasm cooled down, so that it was not until the month of April 1649, that he took in hand the manufacture *of the present clock* (*del presente oriuuolo*) made in accordance with the conception (*concetto*) which his father had already imparted to him in my presence.[2]

Vicenzio Galilei engaged a young locksmith,[3] who had some experience in the construction of large wall clocks. He caused him to make the iron framework, the wheels, and their arbors and pinions, but without cutting them, and he executed the rest of the work with his own hands.

[1] M. Biot's translation is not accurate. The actual words are : "*Finalmente stabilirono il modo che dimostra il quì aggiunto disegno.*"

[2] Here follows a technical description of the mechanism, which M. Biot does not reproduce (see Appendix to this chapter).

[3] Viviani gives his name as Domenico Balestri, and says that he was still living.

Viviani adds: "Vicenzio gave me more than once a demonstration of the connexion of the movement which in this way he had consummated as between the weight and the pendulum, in view of my acquaintance with the invention, and of my having urged him to execute it." [1]

According to the illustration the pendulum was an iron rod carrying a leaden ball which could be made firm at different points of the rod, and it communicated with the wheel work by means of an escapement which might to-day be termed a dead-beat pin escapement.

M. Biot next proceeds to endeavour to rob Galileo of the originality of his pin-wheel escapement, arguing that, because Huygens adopted the universal verge escapement for the application of his pendulum, Galileo must also have adapted his to an escapement already in existence in Italy!

M. Biot asks what could have been Viviani's motive for never divulging his letter, "probably not even to Cardinal Leopold to whom it was addressed!"

After discussing certain reasons that had been put forward, he offers a "very simple explanation. It was a confession. It was remorse for his long neglect, in having realized too late his irreparable fault in permitting to perish in his hands a palm of glory for Galileo and Italy, whilst Huygens' achievement was receiving universal applause. No longer able to revive it, he wished at least to preserve an irrefutable proof that, had Galileo had more zealous and skilful successors of his genius, the glory would have been his. Hence all the minute details in Viviani's letter."

Thus M. Biot.

The discovery by Galileo of the properties of a pendulum

[1] As Viviani's words are here of vital importance, I give them as follows: "*Ciò fatto, volle il Signor Vicenzio che io (come quegli ch'era consapevole di questa invenzione, e che l'avesse stimolato ad effettuarla) vedessi così per prova, e più d'una volta, la congiunta operazione del contrappeso e del pendolo.*"

marked an epoch. It solved the problem which had baffled astronomers, by offering for the first time a means of accurately measuring periods of time. The difficulty was to overcome the manual labour of keeping a pendulum in motion, and of counting the vibrations without error. For this purpose, as M. Biot himself recalls, Galileo, in his celebrated letter to the States-General in 1637, had suggested a connexion between the pendulum rod and a train of wheels, leaving it to the skilled clockmakers of Holland to devise a practical solution. It is true that in that letter Galileo does not speak of a *motive power* to keep the pendulum vibrating ; but it is hardly likely that he would have contemplated the use of *manual* effort as satisfying the requirements of a mechanism to be used at sea for measuring longitudes ! He therefore left this problem to be solved by the clockmakers, to whom the motive power of a falling weight and a coiled spring were perfectly familiar (see p. 85).

In 1639 Galileo published in Paris a work entitled "L'Usage du cadran ou de l'horloge physique universel."[1] In this book, after explaining the movements and properties of a pendulum, its construction and the means for making it swing faster or slower, he deals in chapter iv with its use for ascertaining longitudes and for observations of eclipses. But for this purpose it would be necessary to count the vibrations ; a matter of very difficult accomplishment. "This method was valueless so long as the pendulum was not used for regulating the movement of clockwork."

Huygens was born in 1629. Abandoning the study of the law, he devoted his life to abstruse mathematics and astronomy. His first essay (on the squaring of the circle) was published as early as 1651, two years after the death of Vicenzio Galilei. One of the most pressing problems for a maritime nation was the determination of the longitude

[1] See Père Allexandre, "Traité général des horloges," Paris, 1734, where he gives a résumé (p. 298).

at sea, for the solution of which the States-General of Holland had offered their reward. It has been seen that Galileo had already in 1637 submitted his proposals.

Would it not be a matter for wonder if no inventive genius, whether man of science or clockmaker, had followed along the road which Galileo had pointed out, as leading to the solution of accurate time measurement? An answer, as far as domestic timekeepers go, may perhaps be found in the fact that at that period, and until long after, time was reckoned by solar time, and that clocks, having, with rare exceptions, only a single hand, showed no minutes and still less seconds. The conditions of the early seventeenth century called for none of the exactitude of later days.

But for astronomers the need was ever present, and one must suppose that it was a seconds-counter that primarily interested them, rather than a clock whose equal motion from day to day would cause confusion with the time given by the sundial, to which the world was accustomed.

It would be strange if Huygens' father, when Christiaan was working at the application of the pendulum, had not related his personal acquaintance with Galileo's proposals, and that Galileo's scheme should not have inspired Christiaan in his experiments.

But to return to Monsieur Biot's dissertation. Several observations suggest themselves. The first refers to the letter addressed by Huygens to Cardinal Leopold in 1673. This letter, in view of the fact that Leopold was president of the *Cimento* throughout its existence, and was therefore cognizant of, and even responsible for, the Minutes of 1662 and their publication in 1667, cannot but have placed him in a very delicate position and one of considerable embarrassment.

He is reminded that Huygens had sent him a copy of his "Horologium" of 1658, and of "the benevolence which for a long period Your Highness has deigned to exhibit towards my person and my works"; and "I could have wished," he added, "that you would have intervened to prevent such a thing taking place."

The Cardinal's reply is couched in very guarded terms. He does not attempt to disclaim any share in or knowledge of the entry in the Minutes. All he does is to exonerate Huygens of any acquaintance with Galileo's idea of applying a pendulum to a clock ; or of what had been effected by his son. He acknowledges that Vicenzio had not accomplished much, from which we may infer that he did not place Vicenzio's mechanism on a level with the practical time-pieces of Huygens ; but that is in no sense an admission that Vicenzio had not anticipated Huygens in constructing an instrument in which a pendulum served to control the motion imparted to clockwork. He concludes by saying that "he feels well assured that, if Huygens had had the smallest idea of these particulars, he would not have made any mystery of them."

M. Biot comments on this letter by saying that it is "peremptory, as coming from a person worthy of all respect, cherishing the glory of Galileo, and as well informed as Cardinal Leopold must have been."

In my opinion, as much from what it left unsaid as from what it said, it must have conveyed very poor comfort to Huygens. The priority of Vicenzio's invention is neither expressly nor impliedly abandoned.

Nor must it be forgotten that Galileo's proposals to the States-General distinctly suggest the use of clockwork for mechanical counting, even if they do not allude to a motive power for keeping the pendulum in action. He left it, as he says, to others—the extremely skilful artificers in clockwork in Holland, of whom he speaks—to elaborate a mechanism. When Huygens later on addressed himself to the solution of the problem he must have been acquainted, as I have suggested, with Galileo's proposals—*verbum sapienti sufficit*—and in 1657 he succeeded in accomplishing in a practical manner that which Galileo had suggested in 1637.

Having arrived at a false conclusion, as I submit, in regard to the letters that passed between Huygens and Prince Leopold, M. Biot is equally prejudiced in regard to Viviani's letter to the Prince. The explanation which he gives to account for Viviani not having divulged the letter, namely, his desire to repair his neglect, seems very far-fetched. To support his explanation he assumes that the letter was probably never delivered to Leopold, a gratuitous assumption, as we shall see later. In a footnote he quotes Prof. Albèri as having recognized, after examination of the manuscript, that it was not in Viviani's handwriting, although the alterations and corrections were his. To an unbiased mind I think that the natural inference would have suggested itself, that the very lengthy document was written by an amanuensis and then revised by Viviani. The amended manuscript would thus form the draft of the actual letter written by Viviani himself to Prince Leopold. Viviani would naturally retain the draft of such an important communication. If a fair copy, entirely in Viviani's writing, had been found amongst his papers, there might be some slight justification for questioning whether it had ever been delivered ; but the fact that the document discovered in 1739 was a carefully corrected draft is in itself presumptive evidence, failing proof to the contrary, that it was copied out in its amended form and actually handed to the Prince, *as in fact it was*.

M. Biot next seeks an argument for his theory in the words " *il presente orivuolo* " in the letter. He points out that, as Vicenzio died of fever on May 16, 1649, he only had at the most two and a half months to give to the work. He adds : " We have seen in what a condition he left it. How could Viviani in 1659 (in a letter no doubt intended for posterity) speak of this clock as something *present*, when nine years later it was found still unfinished and abandoned in the experimental stage (*à l'état d'ébauche*) amongst a lot of old furniture ? "

Why does M. Biot jump to the hasty conclusion that the piece found amongst the effects of Vicenzio's widow was identical with the mechanism referred to in the letter? Was it not reasonable to suppose that Vicenzio, in working out this difficult practical problem, would have made more than one trial before he was satisfied as to his success? Viviani actually dwells on the difficulty that Vicenzio experienced in working out a drawing in practical form. The abandoned model in the experimental stage may have been a first attempt, cast aside and forgotten when death suddenly put an end to his labours, but it would constitute a relic which his widow would piously preserve. However this may be, is it to be supposed that the person who made the inventory would have done more than catalogue, as such, an object which to him was "an unfinished iron clock"? It did, however, possess a pendulum! A clock to the lay mind, if it had neither dial nor hand nor case, would naturally be regarded as unfinished; but that is not to say that the clockwork itself with its escapement was not complete. If Vicenzio's object, following on his father's indications, was the construction of an instrument for accurately measuring time for astronomical purposes, the addition of the motion work and index for registering the beats of the pendulum was a mere matter of detail, which any skilled clockmaker could have added in a very short time. In point of fact he aimed at making a complete clock, as will be mentioned later.

The whole point at issue, however, is whether Galileo and his son anticipated Huygens or not, in utilizing the pendulum for controlling the revolution of a train of wheels, and thus securing an accuracy of measurement which the old balance escapement was wholly incapable of giving. It is immaterial to our inquiry whether the motion work should count seconds, hundreds and thousands for astronomical purposes, or record seconds, minutes and hours for ordinary time measurement. Anyone who has seen the working

7

Fig. 15.—Working Model of Galileo's Mechanism.

By courtesy of the Science Museum.

model in action in the Science Museum at South Kensington [1] cannot fail to acknowledge that Galileo's mechanism

[1] It has no motion work, either as a counter or a clock. It was made in 1883, and has been kept going without giving trouble. A few years ago the impulse pallet was renewed. It has a spring motor, whereas Galileo's model was to be driven by a weight.

is far more subtle and ingenious than Huygens' method of controlling the irregular action of the old faulty escapement by the attachment of a pendulum in place of the balance (Fig. 15).

Unless we are to believe, as M. Biot seems plainly to imply, that Viviani's letter was an antedated concoction —a " confession "—to come to light in some uncertain future in order to set Galileo right " in the eyes of posterity," the words " *il presente orivuolo* " refer not, as M. Biot contends, to the unfinished model, but to the *accompanying design of the clock*, as the *pièce justificative* presented to the Cardinal. This is undoubtedly the meaning of the Italian phrase. In proof then of his statements, Viviani sent with his letter to the Prince the design of the instrument devised by Vicenzio for communication to Huygens. Why distort the plain meaning of the words, unless to support an argument, which defeats its own object? What must we think of Viviani trumping up an elaborate story with the object of rehabilitating Galileo in the eyes of posterity, when his sole proof of the perfection of his master's design was an unfinished model?

But if it was the only proof, and if it was " present " when Viviani, after the widow's death, sat down in after-years to write his apocryphal letter, surely he would have carefully stored it with the document for the enlightenment of unborn generations. Such is M. Biot's contention!

There is another point, arising out of Viviani's letter, which M. Biot did not deal with. He concludes that " *il presente orivuolo* " was a fiction, or was represented only by the unfinished model. Was then the circumstantial description given by Viviani of the construction (suppressed by Biot) merely theoretical, or was it based on the practical result obtained by Vicenzio? Galileo and his son in their discussions had fully recognized the difference between theory and practice in mechanical construction. They had between them conceived an entirely new form of escapement, which must have demanded much experimental work for its perfection. When this was accomplished Viviani states that Vicenzio on more than one occasion demonstrated the combined operation of the weight and the pendulum. Clearly then—unless we are to lend no credence to Viviani's statements—the mechanism had been completed, and Viviani's description of it was based upon the model as actually executed. We may even suppose that the technical details, with the drawing, had been furnished by Vicenzio himself, inasmuch as Viviani was not a skilled mechanician.[1]

It must be mentioned that M. Biot—unfairly, as I think—omitted an important and touching passage in Viviani's letter, which tells us that, when Vicenzio and he examined the invention, certain difficulties presented themselves, which Vicenzio hoped to overcome ; he also thought that he could adapt the pendulum to the clock in another way and with other inventions ; but that, having gone so far, he desired to complete it according to the *actual design*, by adding dials for the hours and for the minutes as well ; whereupon he set to work to cut the other toothed wheel (i.e. the motion-wheel). Viviani then proceeds : " But, whilst engaged in this unwonted task, he was overtaken by a most acute attack of fever, and was obliged to leave it unfinished ; and on the twenty-first day of his illness, that is, on May 16, 1649, all the most accurate clocks, together with *this most exact time-measurer*, were by him destroyed (*si guastarono*) and stopped for ever ; whilst he (as it pleases me to believe) passed on, to measure, in the enjoyment of the Divine Essence, the moments of Eternity, that pass all understanding." [2]

[1] In view of the great interest attaching to this description I give a translation of it as an Appendix to this chapter, with a reproduction of the original drawing.

[2] A copy of Viviani's letter and the drawing is given in a tract published in Florence in 1856, entitled " Dell' Orologio a pendolo," from the pen of Prof. Albèri as a supplement to vol. xvi of his edition of the Works of Galileo. This is the " account " referred to on p. 84 (B.M. 8548, D. 34).

Was it not in a moment of delirium that Vicenzio was drawn to this act of destruction? Evidently some overpowering feeling of disgust of all clockwork must have seized him; for not only did his own valued timekeepers perish by his hands, but the conception of his illustrious father, to which all but the finishing touches of the motion work were lacking, was added to the ruin.

The whole facts of the matter, derived from Huygens' correspondence and the Medicean Archives at Florence, showing how and why Viviani's famous Report to Prince Leopold came to be written, may now be stated as follows:

It appears that Boulliau [1] had sent Huygens' "Horologium" to Prince Leopold in October 1658, and in February 1659 he wrote that he was sending the Prince a clock. The Prince replied in March, and, after admitting that Huygens' clock was certainly a beautiful invention, he added that one must not rob Galileo of his glory.[2]

He goes on to state that he had discovered a model (*modello*) already made by Vicenzio Galilei, and that *three years* ago (i.e. in 1656) a clock was made from it by an expert, which he hopes will prove a success when it has been nicely finished (*ridotto al pulito*).

This I take to be the clock made by Treffler, which I have mentioned at the beginning of this chapter. I believe that it has not hitherto been known when he was in the service of the Grand Duke, a point of great importance in this

[1] Ismaël Boulliau (1605–94), author of "Astronomia philolaica," Paris, 1645, close friend and correspondent of Huygens. His work was the first great textbook of Copernican astronomy, to which Dr. Seth Ward replied in 1653.

[2] It should here be noted that the Grand Duke Ferdinand was already in possession of one of Huygens' clocks. In the Register of the "Guardaroba Mediceo" in the Archives at Florence, there is an inventory (N. 959) of the objects in the private Wardrobe of the Grand Duke, dated July 12, 1690. Amongst them was "a clock to attach to the wall with a small pendulum contained in a glazed ebony case with a wavy cornice [23 inches high × $9\frac{1}{8}$ inches wide] with silver hour circle laid on velvet and with a silver label inscribed SALOMON COSTER. It is at present in the saloon, and this clock was sent to the Grand Duke Ferdinand of glorious memory by Sig. Tito Livio Burattini on September 25, 1657, made at The Hague in Holland, and was the first pendulum clock brought into Italy." It no longer exists.

connexion. I owe it to the kindness of Prof. Pietro Pagnini of Florence that I am able to answer this question. He discovered in the Medicean Archives [1] an entry in the Register of the Wardrobe of Ferdinand II, *dated September 2, 1656*, which mentions Treffler as already "the private clockmaker to his Serene Highness." Philip Treffler was a native of Augsburg. When he came to Florence does not appear.

Viviani relates in his Report that Treffler made an instrument for counting the vibrations of a pendulum, and further that he put in order a chamber clock for the Grand Duke giving hours and minutes, and he adds that it was *months later* that the document (i.e. "Horologium") was sent to Prince Leopold.

It is therefore quite possible that he made the first *complete* pendulum clock at Florence, guided by the drawing owned by Viviani and the "unfinished clock" in the possession of Vicenzio Galilei's widow; for no other "model" could have come to the knowledge of the Grand Duke. Treffler would have completed the mechanism by the addition of the motive power, the motion work and the hour circle.[2]

This is confirmed by the relation of Giuseppe Campani, the Roman clockmaker,[3] who states that on a visit to the Grand Duke he was shown "a large old chamber clock which Sig. Galileo had had made." Campani must have misunderstood that the clock was made for Galileo; whereas I suggest that it was the unfinished "model," or Treffler's reconstruction.

The Grand Duke was himself a clock-designer. At this same interview, Campani continues, "His Serene Highness, to my incredible wonderment, showed me a clock of his own

[1] Guardaroba Mediceo (Archivio, No. 609).

[2] An entry in the Register, dated October 10, 1656, states that "Messer Filippo of Augsburg, clockmaker to his Highness," assisted Viviani in conducting important experiments in the measurement of the velocity of sound.

[3] "Discorso di G. C. intorno a' suoi muti oriuoli," Rome, 1660.

invention, which he had had made," and he adds that the Prince was the first to apply the necessary and very simple arrangement that was lacking in Galileo's mechanism for the completion of the clock. But, he concludes, "It must be admitted that Galileo was the first inventor of pendulums and of their application to ordinary clocks."

A letter from Boulliau to Prince Leopold (in Latin) of May 2, 1659, is interesting as showing that at that time Huygens did not claim undisputed priority. He writes that Huygens has warned him, with regard to the pendulum for governing clockwork invented by Galileo, that he does not know who was its first author, and he would take care not to arrogate it to himself only, and thus alone claim the glory. . . . But he thinks that he is entitled to praise for having hit upon the same conception as Galileo. Although he had no other guidance than his own brain, he ought not to boast that he was the first author and inventor of the pendulum.

Boulliau having written to Huygens that he had told the Prince that he (Huygens) was far too honourable to rob another of his fame, the latter replied that, on the assurance of so great a prince, one must believe that Galileo had had the idea before him; did he know the Prince better he would ask for a drawing.

On this being communicated to Leopold, he replied to Boulliau on August 21 that he was sending the design of Galileo's clock, "drawn with the same roughness as is the *model made from it*, which is now in my room." He adds: "You may send it to Huygens, and perhaps next week I shall send you the history of the discovery of the pendulum. I shall also have a drawing made showing how we have fitted the pendulum to our clocks, particularly of a very large striking-clock in the Piazza of our Palace, and shall send it to you."

Such then was the correspondence that led up to the Report, drawn up by Viviani *at the request of the Prince*, and completed on *August* 20, 1659. It was sent by Leopold to Boulliau, together with the drawings of the two clocks.

For some reason it was not until December 19 that Boulliau informed the Prince that he had sent the drawings to Huygens ; and he added that he would have sent a copy of the Report as well, if he had had time for it (*si mihi vacasset*). One of the drawings was the rough copy of the Prince's model, which Huygens received on January 15, 1660, or rather it was a copy of it made by Boulliau.

The drawing in question, made in pencil, measures 13 × 8½ inches, and bears the inscription in Huygens' handwriting : " Missa a Ser° Principi Leopoldo ad Bullialdum, ab illo ad me : R. 15 Jan. 1660, cum descriptio mei Horologii edita fuisset A° 1658," i.e. (" Sent by his Serene Highness Prince Leopold to Boulliau, by him to me : Received January 15, 1660, when the description of my clock had been published in 1658 ") (Fig. 16). And on the drawing is also this note in Boulliau's hand : " Horologe commencé par Galileo Galilei avec vn pendule." Not only that, but there is a second letter from Boulliau, dated January 9, 1660, with which he sends a similar drawing, and he adds : " You may keep the drawings. I have kept the copies sent from Florence. I had yours made after them." [1]

It would appear that the Report—a very lengthy document—never was sent, and consequently Huygens was in ignorance of the whole story. His acknowledgment (in French) of the drawings, dated January 22, is, however, of particular interest, and I therefore quote it in full :

> You have given me great pleasure by sending me the drawing of the clock commenced by Galileo. I see that it has a pendulum, just as mine has, but not applied in the same manner, because, in the first place, he has substituted a much more complicated invention, instead of making use of the wheel which is termed the crown-wheel. In the second place, he has not suspended the pendulum by a thread or narrow band, so that the whole of its

[1] Boulliau was then in Paris.

Fig. 16.—DRAWING OF GALILEO'S CLOCK, SENT TO HUYGENS
(Much Reduced)

By courtesy of the University of Leiden

weight rests on the spindle upon which it moves ; this doubtless is the principal reason why his method was not much of a success; for I know by experiment that its motion becomes far more troublesome, and the clock is liable to stop. Although, then, Galileo had the same idea as myself as regards the use of the pendulum, that is rather in my favour than the contrary ; because I have effected that which he did not succeed in accomplishing ; and in any case, I never had from him nor from anyone whomsoever any hint or suggestion (*acheminement*) as to this invention. If anyone should ever prove the contrary, let him brand me as a plagiarist, thief or whatever else he may fancy.[1]

It will thus be seen that Huygens had failed altogether to comprehend the ingenuity of Galileo's mechanism, with its invention of an entirely novel form of escapement—a complete break with tradition. Huygens, on the other hand, was to the last obsessed with the idea that the problem of an accurate time-measurer was to be solved by controlling the old crown-wheel and verge of the fourteenth century by means of a freely swinging pendulum, in place of the old foliot balance.

He must, however, at that time have felt convinced that in the application of the pendulum he had been forestalled ; for on April 14, 1660, in a further letter to Boulliau he asks : " How can I convince the Prince that neither I nor anyone else in this country [Holland], as far as I know, had ever heard speak of such a thing before I published it ? "

Very different from this humble attitude is that adopted in Huygens' Preface to his " Horologium Oscillatorium " in 1673, dedicated to Louis XIV. After stating that his invention had acquired universal knowledge after the publication of his " Horologium " and the dissemination of his clocks, he proceeds (in Latin) :

[1] It is only fair to recall, as some excuse for Huygens' inability to appreciate Galileo's mechanism, that the drawing sent to him was not reproduced from that discovered by Prof. Albèri, but was a *copy* of the rough drawing made from Prince Leopold's model. A comparison of the two illustrations will show, however, that the difference is not such as to convey a false impression of the principle of the escapement. (See illustration facing Appendix to this chapter.)

It is quite obvious what must be thought of those, who seven years later extolled in their writings the same construction (*sic*), as though it had proceeded from themselves or their friends. As to those who endeavour to ascribe priority to Galileo, if they say that he had attempted though not completed the invention, they would seem to detract more from his merit rather than from mine, inasmuch as I prosecuted the same investigation with better success than he. But when, as a certain learned man has recently alleged, they maintain that the invention was brought to completion either by Galileo himself or by his son, actually exhibiting clocks of this kind, I know not how they can hope to gain credence, seeing that it is scarcely likely that so useful an invention could remain unknown for eight whole years, until it had been brought to light by me. But, if they say that it was intentionally kept secret, they are asserting precisely what anyone can pretend, who seeks to arrogate to himself the priority of an invention. And supposing even that this were proved, it would, even so, in no way affect me, unless at the same time it could be shown that I alone was acquainted with a thing that was kept hidden from everyone else.

It cannot be doubted that the " certain learned man " against whom Huygens directed his pungent remarks was Viviani ; and it is therefore interesting to read his comment upon them, contained in a letter to Count Lorenzo Magalotti—the former Secretary of the Accademia del Cimento—dated July 24, 1673, upon receipt of Huygens' book. He writes :

I fail to understand why " Sig. Ugenio " should exhibit so much passion and be so jealous of the priority in this matter, seeing that anyone who is cognizant of the sublimity of his inventive genius will always and readily believe that he may yet very well have invented subsequently and of his own initiative, and without any previous indication (*motivo*) and information, such application of the pendulum ; in which case it is very certain that he is deserving of the same praise as though no one else had ever thought of it before ; and I myself, after now seeing this marvellous work of his, consider this as being undoubted.

It will thus be seen that Viviani, whilst giving Huygens full credit for independent invention, does not waver in his attribution of the priority to Galileo.

The question may not unnaturally be asked why Galileo's wonderful invention was not turned to account by his friends. To my mind the answer is twofold. Like Huygens, they did not fully appreciate its excellence. Moreover, it was known, as Prince Leopold wrote to Huygens, to very few persons, and those a body of scientists who, unlike Huygens, beside their great attainments as mathematicians, astronomers and physicists, were not endowed with his practical and commercial instincts. The idea of patents and profits never entered into their calculations.

The pity is that nothing of all these priceless "documents" has survived to confirm the written evidence. But in my humble judgment the "old unfinished iron clock *with pendulum*" alone establishes irrefutable proof that Vicenzio Galilei had anticipated Huygens by at least eight years.

As I began the story of this nearly three-hundred-year-old controversy by inviting M. Biot to open the case for Huygens, I will conclude by expressing the opinion that it was deplorable that so distinguished a man of science, in his desire to establish Huygens' claim to priority, and to demolish that put forward for Galileo and his son, should not have hesitated to launch an accusation of gross dishonesty and fraud against Viviani; whilst all the time Prince Leopold's correspondence with Boulliau was available to prove that the charge was as cruel as it was unjustified.

The second drawing, referred to in Prince Leopold's letter to Boulliau of August 20, 1659 (*v. ante*), does not concern the present discussion, but it is of interest as showing how much the application of the pendulum was engaging attention. The large pencil sketch still exists amongst Huygens' papers at Leiden, and is inscribed by Boulliau "Figure de l'horologe a pendule qui est a Florence dans le vieil palais de Medici." I am permitted to reproduce an illustration (much reduced), Fig. 17.[1] It is a strange con-

[1] The original drawing measures 18 × 13⅜ inches.

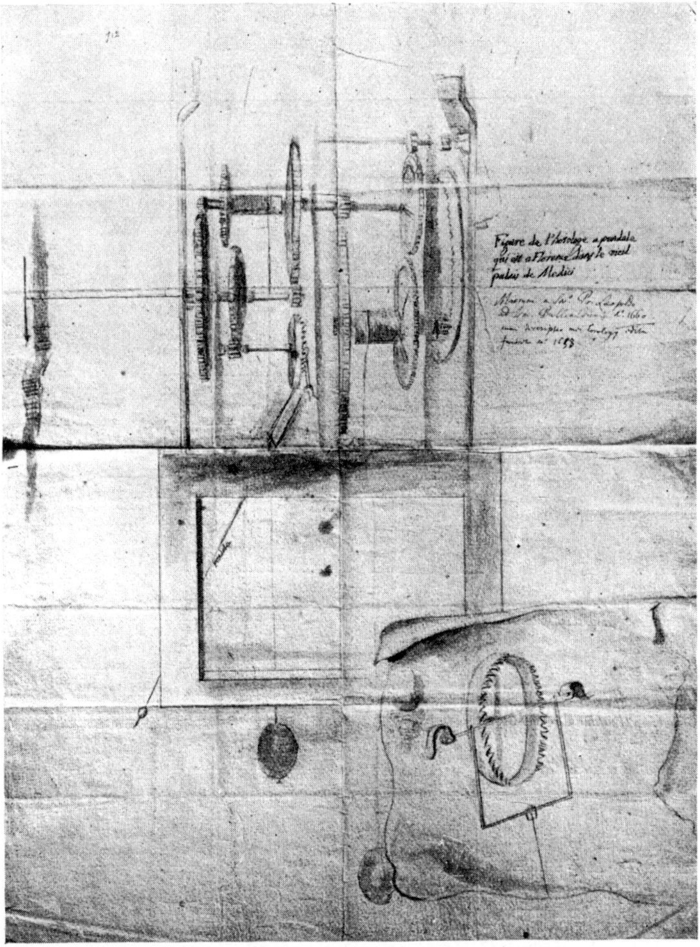

FIG. 17.—DRAWING OF THE MEDICI PALACE CLOCK, SENT TO HUYGENS

By courtesy of the University of Leiden

struction, showing the great wheel at the top and the crown-wheel and verge at the bottom of the movement, and a curious adaptation of the pendulum.

If Viviani's description was accurate and not merely theoretical, it should be possible to reconstruct the mechanism with its novel form of escapement, and thus to establish the genuineness of the claim put forward for Galileo. And this is precisely what took place.

The description, as we have seen, was first given to the world in Nelli's "Life of Galileo" in 1811 ; but apparently van Swinden's lecture had failed to attract attention, and so the matter rested until Galileo's working drawing was discovered by Prof. Albèri.

The subject then excited the interest of M. Boquillon, librarian of the Conservatoire des Arts et Métiers in Paris, and he contributed to the Annals of the Conservatoire a further study of the whole vexed question.[1]

His intention was to rehabilitate Galileo and to endeavour to materialize the particulars given by Viviani ; with the result that he was able to present to the Paris Exhibition of 1855 a model which he believed to realize Galileo's idea. It was some months afterwards, in making a final search amongst Galileo's papers, that the actual drawing came to light, and was reproduced, as has been said, in the supplement to the last volume of Prof. Albèri's work.

The correspondence between M. Boquillon's model and Galileo's design is most remarkable. The only real difference between the two is that M. Boquillon used a star-wheel of twelve points attached to the 'scape-wheel, whereas twelve pins, projecting from the 'scape-wheel, were actually employed. This discrepancy was due to the fact that the word for these pins occurring in the printed version of Viviani's

[1] A short résumé with a commentary by M. Charles Saunier, the well-known horologist, with the comparative drawings was given in the *Revue chronométrique* of October 1861.

letter, was "*pironi*," a word which the Italian dictionaries consulted by him translated as "lever"; hence his star-wheel.[1]

Beside M. Boquillon, Sig. Veladini in Italy had already made an attempt to reproduce Galileo's mechanism from Viviani's letter. In point of fact they might have spared themselves their labour! With Mr. van Swinden's lecture in 1817,[2] to which I have already alluded, he reproduced not only Huygens' earliest design from "Horologium," but also the illustration of Galileo's mechanism which he discovered amongst the vast accumulation of Huygens' manuscripts and correspondence, bequeathed by him to the University of Leiden.

[1] As a matter of fact, Tommaseo and Bellini's great dictionary (1869) gives *pirone* as the equivalent of *perno*, an iron peg or pin. *Pirone* is the technical word for the peg in a harp or piano on which the string is wound. (See Baretti's Dictionary, 1854, and Hoare's, 1915.) Capriglia, in his "Misura di Tempo" (1665), always uses *pirone* for pin.

[2] "Verhandeling over Huygens als uitvinder der slingeruurwerken." The lecture was delivered to the First Class of the Royal Dutch Institute, and appears in Part 3 of its Proceedings for 1817. A somewhat condensed translation appeared in the *Edinburgh Philosophical Journal* for 1821–2, with copies of the drawings.

GALILEO'S ORIGINAL DRAWING (Much Reduced)

From the Medicean Archives at Florence

APPENDIX

The description of Galileo's mechanism contained in Viviani's Report of August 20, 1659

After stating that Vicenzio Galilei engaged Balestri, a young locksmith, to make the iron framework with the wheels, arbors and pinions (but without cutting them), Viviani continues :

The rest of the work he did with his own hands, forming in the upper so-called notched wheel [*i.e.* the escape-wheel] twelve teeth with the same number of pins spaced between the teeth ; and to the pinion on the [same] arbor he gave six [teeth] ; and to the other wheel which drives that above-mentioned he gave ninety [teeth]. He next fixed to one side of the plate which forms the cross-piece [at the top] of the frame the detent or lifting-piece (*scatto*), which acts on the aforesaid upper wheel ; and on the other side he pivoted the pendulum, which was formed of an iron rod, to which was threaded a leaden ball which could be moved by screwing, so that it could be raised or lowered as might be necessary to adjust it with the driving weight [*i.e.* for regulation].

Galilei then explained to Viviani more than once the conjoint operation of the weight and pendulum, as follows :

When the pendulum was at rest it prevented the fall of the weight ; but, when raised and given free play, as it swung past the perpendicular, the longer of the two arms [literally " tails " —*code*] mounted on the pivot of the pendulum raised the detent which acts on and engages with the notched wheel, which latter being pulled by the weight revolved with its upper side towards the pendulum, so that one of its pins pressed on the shorter arm from above, giving it at the commencement of its return an impulse in such a way as to convey a definite motion to the pendulum, causing it to rise to the height from which it started ; so that when it swung back naturally and passed the perpendicular position it once again [by means of the long arm] raised the detent ; the notched wheel then at once, through the force of the weight, resumed its movement, continuing to revolve and to impel the pendulum by the next following pin. And thus in regular action the to-and-fro [motion] of the pendulum was maintained until the weight had run down.

The words in square brackets have been added to make the explanation clearer.

It will be observed that no mention is made of the great wheel, which with its drum, line and weight supplied the motive power ; but these must have been present when the demonstrations were given to Viviani, and we may reasonably infer that he is here speaking of the " most exact time-measurer " which Galilei destroyed in his delirium on May 16, 1649 (see p. 100). The drawing which has come down to us omits these details, and justifies Leopold de' Medici, who had not seen the finished mechanism, in regarding it as a " sketch." The mechanism in the original drawing measures 13⅝ inches.

CHAPTER VII

THE PENDULUM CLOCK

GREAT inventions seldom, if ever, spring fully equipped from the brain of man, like Minerva from the brain of Jove ; but result from the crystallization of ideas that have been floating in the minds of thinkers. When the time is ripe, and the need is urgent, they come to fruition. The science of astronomy had reached so high a state of development that it could no longer be trammelled by methods of approximation, and was impatient at the obstacles placed in its way by the crude time-measurers of the day.

The epoch-making discovery by Galileo of the property of the pendulum could not fail to create a stir, as offering an accurate means of timing observations. Here was found a way of measuring seconds ; but the process was laborious and liable to error in counting, as long as human effort was required. But the germ was sown. The means of solving the mechanical problem was " in the air."

Necessity stimulated invention. But it is not often that the great thinker or mathematical genius can give practical expression to the conceptions which are fructifying in his mind. He needs the help of the trained mechanic, who, by patient experiment in laboratory or workshop, shall determine whether conception can be translated into construction. When many minds are working on the same problem, who shall say who solved it first ? The man who first publishes the solution to the world is acclaimed as the inventor, even though others have in fact anticipated him. Darwin and Wallace, unknown to one another, had simultaneously worked out the doctrine of evolution of species ; but Darwin was first in publication, and his name has been attached to it.

The claim for Galileo and his son that they *first* applied the pendulum to clockwork can hardly be admitted. I have already alluded to the alleged pendulum clock of Justus Borgen at Prague.

Jean Baptiste van Helmont (1577–1644), the eminent Belgian physiologist, states in his essay "De Tempore," contained in his "Ortus Medicinæ," § 51, that he had employed a pendulum for measuring time. The book was published by his son at Leiden in 1648, the home of Huygens, and ran through four editions, the fourth in 1667. He found that the vibrations were always equal in time from the first beat to the last, and that it made no difference whether the weight attached to a cord one foot in length were one pound or several. He makes no pretence to having devised any mechanism for keeping the pendulum in motion ; but his description is evidence that the use of the pendulum for time measurement was in men's minds at a time when Huygens was a boy. From his statement that a pendulum doubled in length gives vibrations of double the duration it is, however, plain that he was not acquainted with Galileo's law, and may have known nothing of the latter's negotiations with the States-General.

On the other hand, we have the positive statement of the renowned German astronomer Hevelius (Johann Hevel, 1611–87) that he had constructed pendulum clocks before Huygens obtained his patent.

In his great work "Machina Cœlestis," published in 1673, he devotes chapter 17 of Book I to the subject under the title "De Horologiis." In this he tells us that Tycho Brahe had vainly attempted to get clocks to keep accurate time, and had been at great expense in experiments. He then relates how between 1630 and 1640 he had tried every known kind of "horologium" of all sizes and shapes, "pensilia, quiescentia et portalia," even such as showed minutes and seconds. But all proved useless for accurate astronomical observation. Then, when he became acquainted with Galileo's experiments in the vibration of a pendulum, he worked between 1640 and 1650, using pendulums of various lengths. The difficulty and labour of counting the beats caused him to devise a counting machine, which he describes. He next endeavoured to find some way of keeping the pendulum in motion by mechanical means, and was in the course of devising a contrivance, which would "not only show the number of the oscillations but also hours and minutes and seconds," when his skilful artificer died. With great difficulty he induced his partner, a Swede, to complete the work. The automaton thus produced remained in continuous use in his house for some time. He then proceeded to the construction of two more pendulum clocks; but his artificer had little time to spare from his astronomical work, and they were not yet completed when Huygens, "*celeberrimus et ingeniosissimus*," invented similar clocks in 1657 and published his description in 1658, "upon which," he says, "I congratulate him most heartily."

He goes on to relate that on the occasion of a visit paid to him in 1660 by the King of Poland, the King was so greatly interested in his clocks that Hevelius presented him with one of them. What was the form of escapement, and what became of the King's clock?[1] To these interesting questions no answer can be given. Possibly the clock may yet come to light.

It is worthy of remark that Hevelius was under no illusion as to the imperfection of Huygens' invention; for he adds that no dependence can be placed on clocks or watches for accurate measurement of time; but he does not despair that in the course of time a more accurate timekeeper may be invented. He would appear to have been ignorant of Vicenzio's mechanism, and at the time of writing he was not aware that the problem had been solved in England by the invention of the anchor escapement.

There is evidence, however, of a still earlier pendulum

[1] Hevelius's observatory and all his instruments were destroyed by fire in 1679, when presumably his own two clocks perished.

clock made at Angoulême about the year 1615. The particulars are given in the "Bulletin et Mémoire de la Société archéologique et historique de la Charente," published at Angoulême in 1892, by M. P. de Fleury, "archiviste de la Charente."

He quotes a letter written on December 13, 1659, by Pierre de Carcavy [1] to Huygens, in the following terms:

> Comme j'aloys hyer faire faire un horloge de vostre jnuention, Je trouvay un honnest'homme d'Angoulesme nommé M. de Boismorand qui m'asseura en auoir un chez luy il y a très longtems à peu prez de la mesme façon, du moins auec pendule, qui fut fait enuiron l'annee 1615 ou 1616 par un Allemand pour feu la Reyne mere de Medicis, qu'elle ne prist point, a cause de son despart d'Angoulesme. Et l'ouurier s'estant marié et decedé quelque tems aprez dans la mesme ville, ledit Seigneur de Boismorand l'a retiré de ses héritiers ; dont J'ay cru vous deuoir donner aduis.

Huygens did not reply until February 26, 1660, regretting that Carcavy had not obtained particulars as to the mode of application of the pendulum, and whether it worked well. He adds: "C'est une chose estrange que personne devant moy n'ait parlé de ces horloges, et qu'a cette heure il s'en decouvre tant d'autres autheurs."

To this Carcavy replied on March 6 that he would have sent a description if the clock had been in Paris, but it is at Angoulême. And there the matter dropped.

M. de Fleury tells us that Boismorand was Garde des Sceaux at Angoulême, and one of the most distinguished connoisseurs and collectors of the town.

Careful research in the archives of Angoulême revealed the fact that the only German clockmaker of the time was one Georg Kloss, who obtained naturalization in 1605. On March 4, 1605, he entered into a five years' agreement

[1] Carcavy, a distinguished mathematician, was one of the original members of the Académie des Sciences, founded in June 1665. He was also a Conseiller au Grand Conseil. He was one of the principal investigators of the calculus, and was in regular correspondence with Huygens, Mersenne and other great mathematicians of the time. He died in 1684.

with the Cathedral Chapter to keep and repair the Cathedral clock ; and on May 26, 1605, he made a similar agreement with the civil authority in regard to the clock in the Church of Saint André. In each case the fee was fixed at 6 livres per annum, new parts to be paid for at market prices.

After 1607 no further mention of him appears in any of the town or Cathedral archives.

It is unfortunate that Carcavy could not personally inspect the clock in Boismorand's collection ; but it would be difficult to refuse credence to the statement of a man, who was not only a " collector " of repute, but also occupied a distinguished official position.

Kloss, moreover, must have been a clockmaker of considerable skill ; otherwise, as an alien, he would hardly have been preferred over native artificers by both the town and the Cathedral body to take charge of their clocks over a period of several years.

Coming to our own country, a claim has been set up for Richard Harris as having constructed a clock with pendulum for the Church of St. Paul's, Covent Garden, in 1642.

The authority for this statement is Thomas Grignion, a clockmaker of repute who died in 1784. His knowledge was derived from his father, also Thomas Grignion, an eminent clockmaker of the first half of the eighteenth century, who lived in the parish and himself made a timepiece for the church. It is fair to assume that he would not have made the statement to his son without good reason.

Thomas Reid [1] remarks that " Inigo Jones, the architect of the church, had twice visited Italy in Galileo's time, and that he might possibly have communicated to Harris what he might have heard about the pendulum."

There is nothing intrinsically impossible in accepting all or any of these claims, once the principles of the pendulum had become known.

[1] " A Treatise on Clock and Watchmaking," Edinburgh, 1826, p. 179.

To Galileo, however, or his son, belongs the honour of devising an *escapement* which was as original as it was ingenious. It was re-invented in 1753 by P. A. Çaron (better known in after-years as Beaumarchais), and developed by Lepaute, as the pin-wheel escapement.

It must not be supposed that anything that I have here set forth on behalf of Galileo and his son is to be deemed as in any sense detracting from Huygens' great achievement; far from it. Whether others had anticipated him in the application of the pendulum or not is really a matter of academic interest only. He was, beyond all question, the first (through Samuel Coster) to put upon the market a clock regulated by a pendulum; and nothing can therefore rob him of the glory of being the pioneer who paved the way for subsequent invention, destined to revolutionize astronomy and navigation as exact sciences; and the debt that the world owes him is not to be measured by any imperfection in his inventions. It is the fate of pioneers to sow for others to reap. The fame of Stephenson is none the less because the locomotive of to-day bears but little resemblance to the "Rocket." And so with Christiaan Huygens.

Unfortunately, his aim was limited, as I have said, to a control of the primitive escapement of the fourteenth century, which had undergone no change, in the belief that the isochronous vibration of a pendulum would overcome its irregular action.[1] To remedy the unequal width of the swing of the pendulum, due to the inherent defect of the escapement and the inconstant force communicated through a train of hand-cut[2] wheels and pinions, however perfect, he applied his discovery of the cycloidal curve traced by a pendulum weight, and endeavoured with his "cycloidal

[1] In his "Mémoires de l'Institut de France" (1808) M. Delambre goes so far as to say that the mere substitution of a pendulum for a balance was not an *invention* at all.

It should here be mentioned that Huygens' application for a *privilège* in France was summarily rejected.

[2] The first machine for cutting wheels is ascribed to Robert Hooke, *ca.* 1670.

cheeks" to compel the pendulum to beat truly. He forgot that the cycloidal curve was performed by a pendulum swinging freely, and independently of the unequal motion of the pallets and crown-wheel. Moreover, the impact of the spring or cord suspension against the cheeks would tend to cause banking.

Nevertheless, the great merits of his invention were the free suspension of the pendulum by means of a cord or steel spring, with the crutch as the means of communication between the pendulum rod and the verge; contrivances which have continued in use ever since in all clocks with the anchor escapement.[1]

It is curious to note in this connexion that the great English masters rarely adopted either the crutch or the separately suspended pendulum in their spring-driven "bracket" clocks, but attached a short pendulum directly to the verge, a system which persisted to the very end of the eighteenth century; nor did the cycloidal cheeks ever find acceptance in this country. Abroad, and particularly in Holland, their use was continued for a time; but there too they were found to be of no practical value and were abandoned.

Huygens' earliest clocks were fitted with a short pendulum beating half-seconds and a light bob, and the cheeks were not yet cycloidal—he did not discover the cycloidal curve until December 1659. It was later on that he adopted as a standard a heavily weighted pendulum beating seconds, as described in "Horologium Oscillatorium," a landmark in the history of horological progress, which by itself entitles him to a niche in the Temple of Fame. But, as appears from a letter to Petit of November 1, 1658, he had already had a seconds pendulum clock made for his astronomical observations.

With his tireless activity he did not confine himself, how-

[1] I shall show later that the idea of the crutch had already occurred to Leonardo da Vinci.

ever, to the problems of the domestic and marine clock; for we find him experimenting upon the turret clock in the church tower at Scheveningen upon a somewhat different principle. He left the crown-wheel and suspended verge in their *vertical* position; but, for the double-armed balance, with its regulating weights, he substituted a rod, which projected on one side only beyond the back of the clock and had a vertical pin fixed to its end.

The pendulum rod was very long, and its suspension swung between cheeks. The connexion between it and the rod was effected by a fork, fixed at about one-third of the length of the pendulum rod, in such a position that the vertical pin was held in the fork. When the weight was applied to the clockwork and the pendulum was set in motion between the cheeks, it was expected to maintain the reciprocal action of the pallets on the crown-wheel in isochronous beat.

Some portions of the escapement, the crown-wheel of eighty-seven teeth, the pendulum and the ornamental suspension bracket for the verge, in the form of a bird, came to light when a new clock was installed in the latter part of the last century. They eventually found a home in the Municipal Museum at The Hague, but have recently been transferred to Huygens' old country seat, Hofwijck, Voorburg.

Huygens evidently found the putting of theory into practice not such a simple matter; for a letter to him, believed to be from Coster, is preserved, under date January 23, 1658, which says: "The clockwork at Scheveningen is ready: it went last night: the bob is a 50-lb. weight, but I think of making it somewhat less: and of altering its spring and chain somewhat: at a guess it has shown an error of a quarter of an hour in fourteen hours."

The clock appears to have been finally completed by Coster, who also "converted" a second clock in the Cathedral at Utrecht in the same year (1658). For the latter work he received 350 guilders, with travelling and living expenses paid, and a gratuity of 80 guilders to his wife (presumably for disturbance).

This form of escapement was adopted in the so-called Zaandam clocks, and subsequently in the common Friesland clocks, with the slight modification that, in place of the projecting rod being held in a fork, it passes through a loop formed in the pendulum. I know of no other application of this system. The cheeks were dispensed with.[1]

Fig. 18.—Zaandam Clock Movement.
In the Science Museum.

[1] A somewhat similar arrangement was, however, perpetuated in the clocks made in Franche Comté in France until the middle of the nineteenth century ; but in these the crown-wheel was set with the teeth pointing downwards,

Fig. 18 illustrates the movement of one of these clocks, which I presented to the Science Museum some years ago.

With the invention of the anchor escapement in England Huygens' system, so far as weight clocks were concerned, became obsolete.[1]

Huygens' patent, it is interesting to recall, stimulated invention in other quarters. Amongst the clockmakers who entered the field the most notable was Simon Douw of Rotterdam. Huygens addressed a complaint to the States-General, to which Douw replied by commencing an action against Coster, as assignee of Huygens' patent.

The result was that both parties were ordered to appear before the Court of Holland, and to produce their mechanisms in the presence of experts. Amongst the latter was Prof. Frans van Schooten,[2] who addressed a letter to Huygens on October 13, 1658, in which the claim of Douw is set forth. Douw, he says, went so far as to allege that it was impossible for Huygens to show how long his pendulum should be in each case !

Douw denied that he had copied Huygens' invention, and stated that he had made use of a shorter swinging rod, and had found it necessary to fix a counterpoise at the upper end of the rod. He employed a vertical staff vibrating round a central point between the two weighted ends.

Huygens, in thanking van Schooten, says that his information " furnishes us with additional arguments for demonstrating the more clearly the sinister proceedings of this plagiarist."

The trial was attended with much expert evidence. Amongst others Benjamin Lisle and Johannes van der Thoorn, clockmakers of Rotterdam, made an affidavit that

and very long pallets were employed. The long pendulum itself swung behind the dial. I have proved them to be excellent time-keepers.

[1] Thus Derham, in the first edition of his " Artificial Clockmaker," 1696, p. 72, speaking of the cycloidal cheeks says : " This way is much ceased, since the Crown-wheel method (to which it is chiefly proper) is swallowed up by the Royal Pendulums."

[2] Eminent mathematician—Huygens was his pupil.

Douw's clocks " differed altogether from the movements of Huygens ; . . . that the motions of the Rotterdam works were based upon another principle," inasmuch as Huygens' invention " consisted merely in the pendulum as employed by Galileo " ; whereas on the contrary under Douw's invention there was not a swinging action, but both ends of the rod, above and below, described a circular turning motion round a central point, " which motions had never been employed by Galileo or any other mathematician."

The " deponents," after describing further particulars of difference and resemblance in the two mechanisms, concluded by stating that the maintaining power adopted by Douw " is not a modification of the invention of the aforesaid Huygens ; but is a new and extraordinary rare invention, which so far as they are aware has never before been employed in any works." They even went so far as to declare that, if they were at liberty to make any such clocks, they would prefer Douw's to that of Huygens.

On the following day an affidavit was lodged by Roussel Brothers of The Hague entirely agreeing with the last affidavit. A certain van Nieuwenhove also made declaration that he had been present at the demonstration, and had examined the clocks and had heard the arguments, and confirmed the above statements in all respects.

The lengthy *procès-verbal* of the case is even more explicit than the above attestations, and arrives at the same conclusion.[1]

The Court eventually delivered a curious decision on December 8, 1658, to the effect that, wherever Douw (who had already obtained an *octrooi* from the States-General) had been granted an " *attache*,"[2] the profits derived from

[1] The whole of the *procès-verbal* of this interesting case is preserved in the Municipal Archives of The Hague, where I have had the privilege of consulting it.

[2] The *octrooi* of the States-General did not confer protection in the different provinces of the Netherlands, unless in each particular province a confirmation —*attache*—was obtained. Douw appears to have secured several ; whereas Huygens would seem to have only obtained one for Gelderland.

both his and Coster's clocks should be shared by Huygens, Coster and Douw !

Douw (born *ca.* 1620) was undoubtedly a clockmaker of repute. Settling early in life in Rotterdam he was appointed clockmaker to the city, where he constructed several public clocks before the days of the pendulum. In 1662 he installed a new clock in the Groote Kerk at Dordrecht. His last great work was the installation of the great carillon of the celebrated maker Jan van Call in the Groote Kerk in Rotterdam, which cost 13,000 gulden. He died on September 9, 1663.[1]

He had originally made overtures to Coster to work with him under his patent, and on this being refused, it was then, according to Huygens, that he managed " by lies and trickery " to get an *octrooi* for himself. A disagreeable incident came to light during the trial. In order to get access to Huygens' work at Scheveningen he endeavoured to bribe the custodian's wife, but without success. It would, however, appear that, after the action was settled, his relations with Huygens were friendly.

The first clocks made on Huygens' system, offered for sale in England, were made by Ahasuerus Fromanteel, as appears from the following advertisement in the *Mercurius Politicus* of October 27–November 3, 1658.

There is lately a way found out for making Clocks that go exact and keep equaller time than any now made without this Regulater (examined and proved before his Highness the Lord Protector, by such Doctors whose knowledge and learning is without exception) and are not subject to alter by change of weather, as others are, and may be made to go a week, or a moneth, or a year, with once winding up, as well as those that are wound up every day, and keep time as well ; and is very excellent for all House Clocks that go either with Springs or Waights : And also Steeple Clocks that are most subject to differ by change of weather. Made by *Ahasuerus Fromanteel*, who made the first that were in ENGLAND. You may have them

[1] These last particulars are derived from an article on Douw by J. J. Moerman in the *Rotterdamsche Jaarboekje* for 1930.

at his house, on the Bank-side in MOSSES-ALLEY, SOUTHWARK, and at the sign of the Maremaid in LOTHBURY, near BARTHOLOMEW-LANE end LONDON." [1]

The Fromanteel family of clockmakers was of Dutch origin. The first Ahasuerus was one of the original members of the Clockmakers' Company in 1631 ; but it was probably his son Ahasuerus, admitted to the Company in 1655, who issued the above advertisement. Members of this family were still engaged in the industry in Holland, and through them their London kinsman was doubtless kept fully informed as to Huygens' invention.

A most interesting discovery has recently been made by Dr. Vollgraff in the Municipal Archives at The Hague : no less than a notarial agreement between Salomon Coster and John Fromanteel, dated September 3, 1657—that is, only eleven weeks after the grant of Coster's *octrooi*—whereby young Fromanteel enters the service of Coster until May 1, 1658, Coster engaging to reveal his secret before expiry of the agreement, and to pay Fromanteel 20 Carolus gulden [2] for each clock made by him, provided that he found his own steel and brass, otherwise 18½ gulden, the clocks so made to remain Coster's property.[3]

[1] The same advertisement appeared in the *Commonwealth Mercury* of November 18–25. Incidentally, this number gives a most interesting account of the last illness of Oliver Cromwell. It is worthy of remark that the spelling of the advertisement is almost entirely modern, unlike the ordinary writing of the period. I found no further advertisements in the various news-sheets in the British Museum up to the end of 1661.

[2] The Carolus gulden was worth approximately 3*s*. 4*d*.

[3] What was the secret to be revealed to John Fromanteel before *expiry* of the agreement? Clearly it must have been something of a different nature from the regular work of making clocks under Coster's patent, for which there was a great demand. In the chapter on the Longitude we shall see that Huygens had finally devised a weight-driven remontoire in August 1664, and that when he communicated his invention to Sir Robert Moray, the latter informed him that he had seen *some five or six years previously* such a clock driven by springs made by Fromanteel, and even a pocket watch. In a lengthy note on this subject by Dr. Vollgraff in vol. xvii of Huygens' "Œuvres," which he is preparing (to be published at the end of 1932), he makes the highly probable suggestion (which he kindly permits me to quote) that this was Coster's "Secret," and that the clock referred to by Moray was one made by Fromanteel in 1658 or 1659. We know that Coster employed a spring motor for "bracket" clocks in 1657, and Huygens, in reply to Moray in October

From this it is evident that no time was lost in sending over one of the younger members of the family (John did not obtain the freedom of the Clockmakers' Company until 1663) to learn the new system.

We may suppose that even before John's return home the Fromanteels were busy with the manufacture of the new clocks from drawings and instructions or models sent over by him, and even clocks to go a month or a year ; because the new design, as the advertisement shows, had already been submitted to experts and approved by Cromwell at some date in 1658 earlier than October. Proceedings were instituted in the same year against Ahasuerus by the Clockmakers' Company on the ground that he was employing more apprentices than the rules permitted, which would suggest that he was unable to cope with the construction of clocks on the new principle without more assistance than the normal requirements of his business would have demanded.

Although the advertisement gives no indication as to the nature of the new " Regulater," it is plain that the pendulum is referred to, and that Fromanteel's work had Huygens' approval in view of the following entries in Evelyn's Diary :

> 1661. 1 April. I din'd with that great mathematician and virtuoso Monsieur Zulichem, inventor of the pendule clock, and discoverer of the phænomenon of Saturn's annulus ; he was elected into our Society.
>
> 3 May. . . . I return'd by Fromantil's the famous clockmaker to see some pendules, Monsieur Zulichem[1] being with us.

The controversy as to Huygens' claims was not limited to the Continent. Our own celebrated Dr. Robert Hooke was in constant dispute as to the Dutch man of science, his

1664, himself acknowledges that he knew that clocks had been made " having two springs, the larger one periodically winding the smaller." Moray states that the rewinding took place half-hourly.

[1] Huygens appears to have been generally known in England by his territorial name, perhaps to distinguish him from his father, of whom Evelyn speaks as " Constantine Hugens, Signor of Zuylichem, that excellent learn'd man, poet and musitian " (June 24, 1671). In his " Journal de Voyage," referring to the second date, Huygens says that he had met " M. Iveling " at Sir R. Moray's house, but no mention is made of the visit to Fromanteel.

attainments as a mathematician and as an inventive genius making him a formidable opponent.

At this point a few words may not be out of place in regard to Hooke's scientific equipment. With none of the advantages of birth and wealth possessed by Huygens and so many of the scientific men of his time, he raised himself to a pinnacle of fame as one of the greatest thinkers and geniuses of an age notably associated with illustrious names; and in England, France and Italy there were giants in those days.

Robert Hooke, the son of a poor country clergyman, was born in 1635, and at the age of eighteen he entered Christ Church, Oxford, in the humble position of "servitor." Already at school he had occupied himself with invention. At Oxford he soon engaged the notice of Dr. Seth Ward (1617–89), the Professor of Astronomy (afterwards Bishop of Salisbury), who instructed him in that science and mathematics.

Thomas Reid[1] states that Ward urged him to try his mechanical genius in contriving an escapement for a pendulum. This he did; and a pendulum clock was used in February 1656 at Oxford for observations of a solar eclipse.

If this can be corroborated, he anticipated Huygens by more than a year; and it would not be surprising if thenceforth he devoted himself particularly to improvements in the faulty timepieces of that age. His talents had already been utilized by Robert Boyle in the construction of his air-pump. Thus, without influence, he was drawn of his own merit into the circle of the illustrious men who, first at Oxford, and then in London, formed the society of scientists which in 1662 received the Royal Charter of Incorporation. In the same year he became its Curator; in 1663 he was elected as a Fellow, whilst in 1677, on the death of Oldenburg, he succeeded him as Secretary. He died in 1703.

John Aubrey (1626–97), elected F.R.S. in the same year

[1] *Op. cit.*, p. 184.

as Hooke, and therefore well fitted to describe him, draws a picture in his " Lives " of an outward appearance far from prepossessing. This physical misfortune, combined with an irascible, jealous and suspicious temperament and a bitter tongue, did not make for popularity ; but his genius could not be denied. It is difficult to avoid the feeling that his noble colleagues, Lord Brouncker, the President, Lord Kincardine, the Hon. Robert Boyle and the rest endeavoured to exploit his poverty and his genius for their own advantage. With Oldenburg he was in constant feud; but of this hereafter.[1]

Amongst his other horological inventions he claimed that of the circular pendulum, referred to in the Transactions of the Royal Society between February and April 1667, and also the spiral balance spring, both of which were also claimed by Huygens. It is not to the present purpose to deal with the latter controversy, to which I devote a later chapter.

He also contributed to the accuracy of timekeepers by the invention of the first cutting machines for clock-wheels.

In further testimony to his inventive genius a passage in Evelyn's Diary of August 4, 1665, is worth quoting :

> I call'd at Durdans, where I found Dr. Wilkins, Sir Wm. Petty, and Mr. Hooke, contriving chariots, new rigging for ships, a wheele for one to run races in, and other mechanical inventions; perhaps three such persons together were not to be found elsewhere in Europe, for parts and ingenuity.

The Transactions of the Royal Society abound in references to Hooke's proposals and inventions of such varied description that it is small wonder that he found no time to bring them to fruition, and that his claims to originality were afterwards contested for want of more substantial proof than his own assertion. His claim to the invention of the anchor escapement is dealt with in Chapter VIII.

[1] In this connexion Oldenburg's appreciation of Hooke, contained in a letter in French to Huygens of June 21, 1675, is interesting. He says : " He is a man of quite peculiar temperament, which one has to bear with all the more patiently, because he possesses a great fertility of intellect in the invention of novelties. *Nullum magnum ingenium sine mixtura dementiæ fuit.*"

CHAPTER VIII

THE "ANCHOR" ESCAPEMENT

THE substitution by Huygens of a pendulum for the foliot balance, even with the cycloidal cheeks, was a failure from the point of view of exact observation. But he had succeeded in producing a far more efficient timepiece than any that were previously in use. The radical defect lay in the retention of the old verge and crown-wheel escapement. The practical utility of the pendulum as a controller of the clockwork was established; and the next step in advance depended, therefore, on the abandonment of tradition, and the discovery of some wholly novel form of escapement. The pin-wheel escapement of Galileo was not divulged to the world, and the new invention proceeded upon wholly different lines.

The vertical crown-wheel of the old balance clock was replaced by a vertical flat-wheel with saw-shaped teeth, controlled by pallets at either end of a curved steel device, suggesting the head of an anchor. Outside the back plate of the movement the prolongation of the verge or arbor which carried the "anchor" was provided with Huygens' crutch, in the fork of which swung a long pendulum, suspended by a steel spring from a bracket, fixed to the top of the back plate. Such was the invention which with little modification has continued in use in ordinary long-case clocks to the present day, an invention destined to revolutionize time measurement.

It is no part of my purpose to chronicle any of the subsequent improvements in the escapement itself or in the pendulum, but only to deal with its origin.

Unfortunately, there is some element of doubt as to its authorship. It has been alleged by Reid (p. 184), Wood (p. 298) and others, so as almost to have become an article of faith, that the first long pendulum clock with the new escapement was made by William Clement in London about the year 1680. I do not know where this date originated.

Derham,[1] however, says that—

> Dr. *Hook* denies Mr. *Clement* to have invented this ; and says that it is his invention, and that he caused a piece of this nature to be made, which he showed before the *R. Society*, soon after the Fire of *London*.

It has been stated that Hooke invented the anchor escapement in 1666. I have not been able to find any proof of this in the records of the Royal Society. There is an entry of that year on August 29, when " Mr. Hooke produced a new piece of Watch-work of his contrivance, serving to measure time exactly both by sea and land ; of which he was ordered to bring in the description." No further reference is made as to his having done so. The contrivance was most probably one of his many experiments with a balance spring. In any case, he would never have proposed a pendulum clock for use at sea, having already, on March 15, 1665, condemned Huygens' attempts.

Frequent entries occur in 1666 to Hooke's horological activities, but most of them refer to " a new contrivance of a circular pendulum applicable to a watch [i.e. clock], and moving without any noise and in continued and even motion without any jerks," exhibited by him on June 13. He was constantly reminded, up to April 18, 1667, to produce a demonstration in proof of his claims ; but from that date no more is heard of it, until Huygens claimed the invention as his.[2]

Smith[3] tells us that " several eminent and ingenious Workmen applyed themselves to rectify some Defects which as yet was found " in Huygens' clock, and he adds that—

> That eminent and well known Artist, Mr. *William Clement*, had at last the good fortune to give it the finishing Stroke, he

[1] Derham, " The Artificial Clockmaker," 1696, p. 96.
[2] Hooke deals with it in his " Lectiones Cutlerianæ."
[3] " Horological Disquisitions," 1694, p. 3, by John Smith.

being indeed the real Contriver of the curious kind of long Pendulum, which is at this Day so universally in use among us.

The general consensus of opinion has, however, accepted the validity of Dr. Hooke's claim to the invention. The entire dislocation of business caused by the Great Fire, coupled with the destruction of most of the clockmakers' premises, must have put a stop for some considerable time to the development of new inventions; and it would therefore not be surprising if Hooke's anchor pallets, if he really invented them, were not generally adopted for some years. If the new escapement was not patented, as it does not appear to have been, there would be nothing to prevent Clement or any other clockmaker from adopting it.[1] Hooke certainly associated himself with Thomas Tompion in the execution of his inventions. There came to light in 1826 a superb year-clock movement, of exquisite workmanship, with the anchor escapement, which bears the inscription:

Sr. Jonas Moore caused this movement with great care
to be thus made A° 1676 by Tho Tompion.

The dial is in the Dutch style, covered with red velvet, and the delicate hands are gilt, indicating that the Dutch influence was still dominant. It has fortunately passed into the possession of the British Museum. The case is of much later date.[2]

We have, however, positive evidence that Clement actually employed the anchor escapement and long pendulum in a *turret* clock, in the shape of a very interesting specimen now exhibited in the Science Museum, South Kensington.

[1] It is remarkable that there is no record in the Patent Office of any horological patent being granted between March 1665 and March 1694, a period during which our great clockmakers were particularly fertile in invention.

[2] Wood refers to it in his "Curiosities of Clocks and Watches," 1866, p. 111; but the record that he quotes that it was completed when the Plague broke out, and that in the confusion it was placed without a case in the lumber room of the Royal Society's house, is ridiculous, seeing that the Plague took place in 1665.

This clock was made for King's College, Cambridge, and was thence transferred to St. Giles's Church, Cambridge, and in 1926 was presented to the Museum. It bears the inscription : " Guglielmus Clement Londini fecit 1671."

There is, moreover, John Smith's own evidence that domestic clocks with long pendulums—known as " royal pendulums "—were in common use in 1675, when he published his " Horological Dialogues," as the following passages show :

> Those Clocks, who have their motion regulated by a Pendulum, are more excellent than those who are regulated by a Ballance, and those that are regulated by a long Pendulum are far more excellent than those that are regulated by a short one. [P. 18.]

On p. 34 he speaks of " setting up long swing Pendulums after you have taken it from the coffin."

On p. 47 he says that whilst he has " nothing more to deliver concerning these ordinary ones [Pendulums], I confess something more might be delivered concerning the more richer kinds."

And on p. 54, speaking of the wear of clocks for want of oiling, he says : " This I fear will be the fate of those long swing eight day Pendulums *so much at this day celebrated*."

From this it is abundantly clear that a long pendulum was in general use for house clocks prior to 1675, although it is not absolutely certain that Smith is referring to it in connexion with the anchor escapement, bearing in mind Huygens' illustration in his " Horologium Oscillatorium " of 1673 of a seconds pendulum applied to the old-fashioned verge, and that he had used it himself in 1658.

In Derham's " Artificial Clockmaker," even at the date of the first edition in 1696, the directions given for converting a balance clock into a pendulum clock are for the substitution of a contrate-wheel in place of the old vertical crown-wheel, and planting the new crown-wheel above the top plate, with the pinion on its vertical arbor engaging

Fig. 19.—English Lantern Clock with Circular Balance, ca. 1620.

In the Author's collection. From "The Dictionary of English Furniture." By courtesy of "Country Life."

in the contrate-wheel ; in fact, Huygens' method. How much simpler was it to replace the old crown-wheel by the new vertical 'scape-wheel, with the verge carrying the anchor set above the top plate, a hole being made in the plate to allow of the pallets engaging in the 'scape-wheel.

The only domestic clock in common use in England before the application of the pendulum was the so-called "lantern" clock to hang against the wall, with its brass cage, surmounted by a dome containing a large silvered bell, below which swung the *circular* balance. The "foliot," or bar balance, was not used in the English domestic clocks. Such clocks were very inaccurate timekeepers. Fig. 19 shows one of the rare survivals in my collection.

John Smith, in his "Horological Dialogues" (1675), says of them :

> These movements going with weights must be brought to keep true time by adding to or diminishing from them : if they go too slow you must add thin shifts of lead to the weights to make it go faster, but if it go too fast then you must diminish the weight to make it go slower, so that, whensoever you find it either to gain or loose, you must thus, by adding or diminishing, rectifie its motion : note that these Ballance movements are exceedingly subject to be altered by the change of weather, and therefore are most commonly very troublesome to keep to a true time. (P. 43.)

It is small wonder, then, if everyone who could afford it had his lantern clock converted to a pendulum clock ; hence the great rarity of "unconverted" specimens.

Whilst I cannot affirm that Huygens' verge escapement was never adopted, I have not in my own experience come across, or heard of, a converted clock with a *long* pendulum that had not been fitted with the anchor pallets. These converted clocks have pendulums of varying length. As such clocks were only furnished with an hour hand, the necessity for a pendulum beating seconds was not apparent. Even the earliest types of the new-style clock, enclosed in a "long case," were only provided with an hour hand and a minute hand. But it was soon perceived that, without adding to the motion work, it was easy to mark the seconds by prolonging the arbor of the 'scape-wheel through the dial and fixing to it a small hand revolving in a subsidiary circle graduated from 0 to 60. For this purpose all that was necessary was to calculate the going train, so that the 'scape-wheel in combination with a pendulum of suitable length should turn once in a minute. This soon became the standard system; but early experiments were made with a much longer pendulum, giving a beat of one and a quarter seconds. Such clocks were notably made by William Clement; and, if he was the first to introduce them, it is possible that they were the origin of the belief that the anchor escapement itself was due to his invention.

So far I have been dealing with this escapement as applied to weight-driven clocks. We should expect that such an improvement in timekeeping would also have been employed in spring-driven bracket clocks with a short pendulum; but for some reason, for which I have failed to find any explanation, the general practice, in England at any rate, up to quite the end of the eighteenth century, was to perpetuate the old verge escapement, without either the suspended pendulum or the crutch. Even George Graham, the inventor of the "dead-beat" escapement about 1715, which marked such an advance on the old anchor pallets, made bracket clocks with the verge.

I cannot leave the subject of the pendulum without mention of the fact that the idea of controlling the verge escapement by a rod attached to the verge at right angles, with a bob at its extremity, had already suggested itself to that universal genius, Leonardo da Vinci.

In the Biblioteca Ambrosiana at Milan there exists a large quantity of single sheets of various sizes in Leonardo's handwriting, which were bound up, without indication of date, into one volume, known as the "Codice Atlantico." In any case, they were written during the last years of the fifteenth and the early part of the sixteenth centuries. They contain numerous rough sketches illustrating mechanical devices, but unaccompanied in some cases by any descriptive text. Amongst these there occur on folio 257 Ra the drawings which are here reproduced (Fig. 20).

The upper sketch shows a wheel or pinion driving a horizontal crown-wheel by means of pins set in its periphery. Above the crown-wheel is shown a verge with pallets, terminating in a pivot, with the weighted pendulum rod fixed to it vertically. Apart from the method of driving, this is precisely the form of the verge escapement, connected with a short pendulum and bob, which is found in English bracket clocks up to the close of the eighteenth century. The wheel at the top may be intended for a circular balance to act as a regulator; whilst the barrel with the coiled rope below may be meant to indicate the method by which a train of wheels to connect with the escapement was to be driven. The text, in Leonardo's "mirror" writing, refers to the coiling and uncoiling of the rope, but unfortunately gives no explanation of the escapement and the means by which its action was to be controlled.

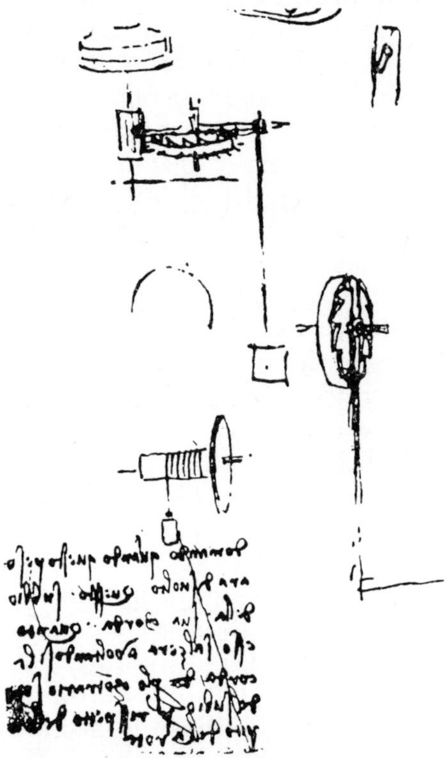

Fig. 20.—Leonardo da Vinci's Pendulum.

The lower sketch seems to contain the germ of the crutch.

As the manuscripts were deposited in the Biblioteca Ambrosiana in 1637, it is quite possible that these sketches became known to Italian clockmakers. Mr. Percy Webster—than whom no one has had greater opportunity for forming an opinion—informs me that, from specimens that have

Fig. 21.—Pendulum Clock, by Camerini, dated 1656.
By courtesy of Mr. Percy Webster

passed through his hands, he is convinced that in the early seventeenth century small clocks were made with a light pendulum and bob without a crutch, and that they had not been subjected to subsequent "conversion."

A beautiful little clock by Camerini, answering to this description, in Mr. Webster's collection is here illustrated (Fig. 21). Below the hour circle is inscribed on a scroll the date AN. DNI 1656, that is to say, a year prior to Huygens' patent.

M. Frémont, in his treatise, which I have cited elsewhere, reproduces other sketches from Leonardo's manuscripts illustrating the alternating action of a heavily weighted pendulum for operating double-plunger pumps. Other instances of its use for various mechanical processes are also given, taken from engravings in Besson's "Théâtre des instruments mathématiques et mécaniques," published at Lyons in 1578.

It is therefore certain that as a constituent part of machinery the pendulum was regularly utilized long before its properties were investigated by Galileo.

We must, however, regard these devices as a method of exerting, and not as a means of *controlling* power already derived from some other source.

CHRISTIAAN HUYGENS (Aged 42)

From the painting by Caspar Netscher. By courtesy of the Gemeente Museum, The Hague

CHAPTER IX

HUYGENS AND THE DETERMINATION OF LONGITUDE

Huygens, as we have seen, conceived the idea of adapting a pendulum to clockwork in lieu of the old bar balance, whose action was erratic, and therefore useless for accurate time measurement. This was in December 1656, and in June 1657 his clock was completed, and was granted a patent (*octrooi*) by the States-General of the United Provinces of the Netherlands, as already described in Chapter VI.

Whether at this time Huygens only had in view the use of his clock for general purposes and for astronomical observation, or whether, as Dr. Hooke alleges, its use for determining the longitude was first suggested to him by Alexander Bruce, I am not prepared to say; but certain it is that Bruce was the first to put the matter to practical test. At any rate, when Huygens published the description of his clock in 1658 ("Horologium"), he refers to the question of the longitude, leaving it for future investigation by himself or others.[1] There is, however, no pretence that the clock therein described and illustrated would serve that purpose.

Alexander Bruce (16 ? –80), always spoken of as "Brus" by Huygens—succeeded his brother in 1663 as second Earl of Kincardine. He was an accomplished linguist, well informed in medicine and science, and with a great talent for mechanics, the possessor of an extensive estate and mines in Scotland, which brought in a large revenue, and enabled him to lend pecuniary assistance to the exiled Charles II. Political reasons during the Commonwealth obliged him

[1] *Ut jam de Longitudinum, quam vocant, scientia dicere omittam: quæ si unquam extitura est, desideratumque tantopere usum cursui navigantium præbitura, non aliter, quam vectis per mare exquisitissimis atque omni errore vacuis horologiis, id obtineri posse, multi nobiscum existimant. Verum hæc res vel ipsi mihi, vel aliis quandoque curæ erit.*

to leave Scotland in 1657 ; and, after a stay at Hamburg and Bremen, he took up his residence at The Hague, during which he married a wealthy Dutch wife, daughter of Cornelis van Aerssen, on June 16, 1659. Having thus entered Dutch society, we may suppose that he became acquainted with the distinguished Huygens family, Sir Constantijn (knighted by James I) and his four sons, of whom Christiaan was the second.

Shortly after the Restoration Bruce returned to England ; and it appears from Huygens' Diary that he visited Bruce in London early in 1661, when the pressing problem of the longitude and its solution by a clock was doubtless discussed between them. It was on the occasion of this visit that, as Evelyn records, " that great mathematician and virtuoso Monsieur Zulichem, inventor of the pendule clock, and discoverer of the phænomenon of Saturn's annulus was elected into our Society " (April 1/11, 1661).[1] Bruce was also one of its earliest members.

As a consequence of the anti-Puritan agitation in England, and in spite of his services to the exiled King, Bruce, as a staunch Presbyterian, found it advisable to return to Scotland later in the year, and at some time towards the middle of 1662 he was once more in Holland. It was now that he applied his mechanical genius to the adaptation of Huygens' clock for the purpose of determining the longitude at sea. The first mention of his activities occurs in a letter of Huygens to his brother of November 9, 1662, referring to a trial to be made on a voyage to Scotland, which, however, did not take place.

Much of the correspondence on the subject took place with Sir Robert Moray.[2] Writing to him on December 1, 1662, Huygens states that " Bruce has had two clocks made

[1] The Charter incorporating " The Royal Society " was granted in August 1662.

[2] Sir Robert Moray was the first President of the original Society in London, and was succeeded by Lord Brouncker (? 1620–84) when the Royal Society received its Charter.

. . . as to their maintaining accuracy at sea we can only judge by experiment."[1]

From the description of his clock in " Horologium " it is evident that Huygens was thinking more of astronomy than navigation. This is not surprising, when we remember that in 1655 he had been working with his brother Constantijn on the formation of lenses, with the aid of which in a ten-foot telescope he made the wonderful discovery of Saturn's ring.

His clock was weight-driven, with a "maintaining power" consisting of an endless cord passing over two pulleys and kept taut by a counterpoise, so that the driving weight still exercised its pull on the great wheel during re-winding, and prevented any loss of time during the process. This was Huygens' first invention of a maintaining power. The returned end of the crutch had a hole drilled in it, through which the pendulum rod was passed. The object of this, as he explains, was to keep the crutch in position in the event of any violent movement. The length of the pendulum was calculated for beating half-seconds, and was $\frac{10}{12}$ of a Rhenish foot (equal approximately to $9\frac{3}{4}$ inches ; *see* Fig. 14).

In further letters up to the end of December Huygens invariably speaks of " his [Bruce's] two marine clocks," and he adds that they were in dispute as to the division of the profits, as Bruce maintains that he has contributed not a little by " his invention " in the adjustment of the pendulum so as to resist the motion of the ship.

Eventually Bruce crossed to England at the beginning of January 1663, but in such rough weather that one clock was thrown down from where it was suspended and the other so much shaken that it stopped.

Hooke's recollection, after ten years' interval, was apparently at fault when he gives February or March 1664

[1] The correspondence from which I quote is contained in the monumental " Œuvres Complètes " of Huygens, who left all his papers to the University of Leiden. The publication, commenced in 1888, is still continuing. The letters are mostly written in French.

as the date of a fresh trial in the King's pleasure yachts ; for Sir Robert Moray, writing to Huygens on February 19, 1663, says " next week we purpose making a little trip at sea as far as the Dunes to try *Mr. Bruce's clocks.*" Hooke, in his account of this trial, contradicts Huygens, affirming that one clock was made in England (by Fromanteel). This is quite likely, seeing that the clock, which was thrown down and no doubt much knocked about in the heavy sea, must have sustained much damage. But for comparison the English clock must have been of the same design.

The agreement proposed by Bruce as to profits, if the " invention of the longitudes " was successful, was to be equal shares, a division which Huygens thought inequitable, as his clock was the basis of the invention ; but he did not demur, so as to avoid a dispute. Later on a disagreeable incident occurred. In January 1664, Kincardine wrote to Huygens as if *he* alone were the inventor of the " longitudes," and as if he were to give Huygens his share as alms.[1] On January 9, in a dignified reply, Huygens concludes by saying that rather than accept such conditions he would for his honour's sake abandon the whole invention to Kincardine.

These differences were afterwards composed by Sir Robert Moray's intervention. Meantime several letters had passed between Moray and Huygens in November and December 1663, as to taking out a patent. Kincardine at Moray's instigation agreed that it should be in the name of the Royal Society, and Moray suggested that the Treasurer, Abraham Hill, should make the application. This was agreed to, and the patent was finally entered (with inventions of other persons) on March 3, 1665.[2]

[1] After Holmes' return from Lisbon, as related later.

[2] The Letters Patent, dated March 3, 1665, at the Record Office read as follows :

HILL, Abraham.

A new way of makeing of watches and clocks to bee vsed at sea for exact measuring of tyme, towardes the finding the longitude and knowing the true course and place of a shipp, differenced from all other sorte of watches by

What then was Bruce's invention, apart from what Hooke later on described, in a paper read before the Royal Society, as a " case made very heavy with lead and suspended underneath the deck by a ball and socket of brass," in fact, some arrangement of gimbals as is used for a compass or chronometer? That his clocks were driven by a spring and not by a weight, which latter would have been impracticable at sea, and also that the train, unlike Huygens', was calculated for a 6-inch pendulum (Huygens says in one letter 7-inch) emerges from the documents. But already in " Horologium," 1658, the spring motor is mentioned, and Coster's clocks, as we have seen, were thus actuated in 1657.

The one point of originality in the mechanism that appears in the correspondence is in the formation of the crutch, which is described by Huygens in " Horologium Oscillatorium " as resembling an inverted F. Fortunately a drawing of it has recently come to light in Huygens' MS. Book " B," stuck in at a place which would indicate August or September 1662. The illustration (Fig. 22) clearly shows the suspension of the pendulum between the " cycloidal cheeks," with its rod held between *two* jaws or " catches " in the crutch. The plate which carries the pendulum measures $7\frac{7}{8}$ inches in the original drawing.

The idea was ingenious, the object being to prevent a rotary motion of the pendulum at sea; but in practice it caused " banking," and increased friction. The design in " Horologium Oscillatorium " shows that Huygens had discarded it in a domestic clock, and adopted the single-forked crutch, which was then, and has been ever since in use.

having, instead of a ballance, a rodd of wyer, or a thynn narrow plate, with a weight at the lower end thereof, called a pendulum, and at the vpper end an arme with twoe catches or holes to moue it, and certaine crooked places (*plates*) or cheekes for regulating the motion thereof, which motion is p'duced by one or more springs or weights, the said watches being fitted with balls and socketts to hang by for goeing steadily at sea. (No specification was enrolled.)

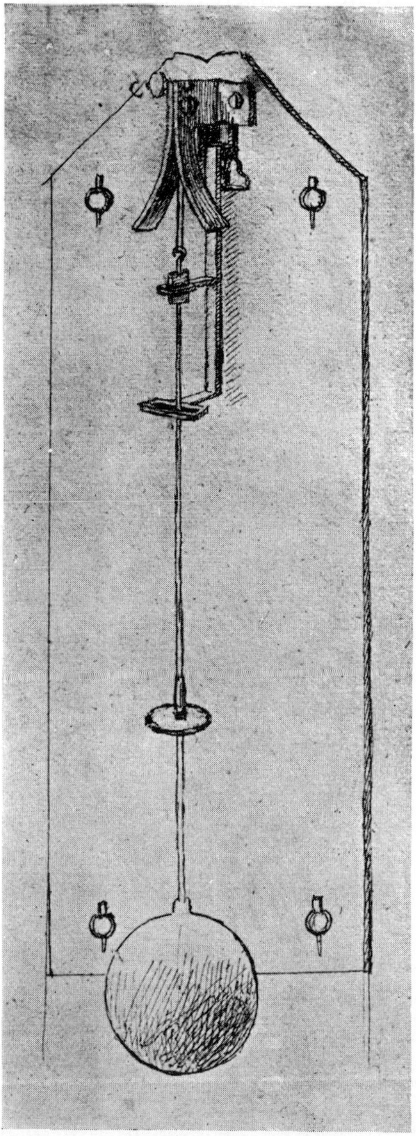

Fig. 22.—Bruce's Double-fork Crutch.

A report on the behaviour of these clocks —"Bruce's clocks"— during a voyage to Lisbon between April 29 and September 4, 1663, is preserved in the British Museum (Sloan 598). The clocks were of different sizes, the larger of which went with fair regularity; but the smaller frequently stopped and had to be taken to pieces and oiled during the voyage.[1]

Huygens was, however, satisfied with the result; for, in a letter to Moray of November 11, he says that he was greatly pleased, not having dared to hope that the clocks would have been so successful. And again, after receiving the Captain's journal, he writes on November 18 that it is a great point to know that a

[1] The clocks were placed in the charge of Captain—afterwards Admiral—Sir Robert Holmes.

well-made clock, as *that of The Hague*, can withstand the worst storms without stopping. He adds that the variations were not every day the same, as he had also observed *with these same clocks* when Lord Kincardine and he were busy adjusting them, and he was glad to see that they went as well at sea as in his room. From a letter to his brother of November 30 it appears that the larger clock was made by Severijn (Oosterwijk) ; and he explains that the other stopped occasionally because the pendulum bob touched the case.

A further trial of " Bruce's clocks " was made by Captain Holmes in 1664 on an expedition which was fitted out to harry the Dutch on the West Coast of Africa and in America. On this occasion he seized the Dutch forts in Guinea in February, although war had not been declared, and he then proceeded to capture New Amsterdam (New York) in August, after which he sailed to the West Indies, finally reaching Plymouth at the beginning of December. It was during this last part of the voyage that the clocks were considered to have given a good account of themselves.

Holmes sailed back across the Atlantic, and arrived at " Cape Corso," now known as Cape Coast Castle. Why he did so does not appear ; but it may be that he wanted to ascertain how matters stood on the Guinea Coast after his long absence. He then touched at the Island of St. Thomas, just above the Equator, where he adjusted the clocks, and then set a course westward for some 700 leagues [1] without changing direction to obtain a favourable wind to carry them back to the African coast. When they had sailed NNE, about 200 to 300 leagues,[1] the captains of the three other ships under Holmes' command stated that in their judgment they were far from the African coast, and that their water would not hold out until they reached it, and they advised returning to Barbados. Holmes thereupon summoned a council ; and an examination of their logs as compared with his own reckoning showed errors of 80, 100,

[1] These are the *corrected* distances, as mentioned below.

and 120 leagues; whereas, according to his clocks they were within 30 leagues of Fuego, one of the Cape Verd Islands, where they could get water ; if at the end of two days his calculation was wrong he would supply them with water, of which he had an ample supply, and make for Barbados. On the following morning Fuego was sighted.

A report was written by Moray to Huygens on January 23, 1665, after Holmes had arrived in London.[1] Huygens was naturally delighted with this report ; and on February 5 he writes to his friend Chapelain in Paris that " the usefulness of the clocks surpasses my expectations. I had not imagined that clocks of *this first construction* would succeed so well, and I had reserved my chief hope for the *new ones*."

The distances given by Moray having been called in question by members of the Royal Society, Pepys was deputed to visit Holmes, whom he failed to see, but on March 15 he reported as follows :

> Mr. Pepys gave an account of what information he had received from the master of the Jersey ship, who had been in Company of Major Holmes in the Guinea voyage, concerning the pendulum watches, viz. that the said master affirmed, that the vulgar reckoning proved as near as that of the watches, which, added he, had varied from one another unequally, sometimes backwards, sometimes forward, to 4, 6, 7, 3, 5 minutes ; as also that they had been corrected by the usual account. And as to the island, at which they had watered, the said master declared, that it was not Fuego, but another 30 miles distant from the same westward.[2]

Holmes was under arrest in the Tower, which may account for Pepys obtaining his information from the ship's master.[3] Mr. Hooke thereupon gave his reasons for considering " that no certainty could be had from these watches for the longitudes."

[1] A translation of this letter into Latin is included in the " Opera Varia," 1724, vol. i, pp. 211–12.

[2] Birch's " History of the Royal Society," vol. ii, p. 23.

[3] The corrected distances are given in " Horologium Oscillatorium," p. 47, as also in vol. i of the Philosophical Transactions. (Moray replied that they had made Fuego, but proceeded for watering.)

This filibustering expedition of Holmes to the Guinea Coast was planned by the Royal African Company, the patron of which was the Duke of York, with Holmes as the eventual scapegoat. The entries in Pepys' Diary on the subject are amusing. On May 29, 1664, he has a discourse with Mr. Coventry about "the Warr with the Dutch . . . and what Holmes without any commission hath done in taking an island and two forts . . . and he [Coventry] believes that Holmes will have been much puffed up with this." And on January 9, 1665, he writes : "Holmes was this day sent to the Tower ; but I perceive it is made a matter of jest only." And so it proved ; for Holmes cleared himself to the King's satisfaction.

In any case, this is the last we hear of English trials with "Bruce's clocks," and of Holmes in an attempt to determine the longitude at sea.

This brings us to a very interesting episode. When the illustration of Bruce's crutch (Fig. 22) was removed from the MS. book there was found covered over by it a rough sketch of a remontoire device[1]; which I am also permitted to reproduce (Fig. 23).

This drawing (measuring 6⅜ inches) was no doubt made about the same time, namely, in August or September 1662. It follows therefore that Huygens was already pondering

[1] The principle of the remontoire is the application of power directly to the escapement independently of the main clockwork, which only serves to restore the power at intervals. The idea is ingenious, but in practice the friction on the escapement is fatal to success. Lieut.-Commander Gould describes it as the *ignis fatuus* of all the great makers. Reid states (I know not on what authority) that it had been attempted by a German clockmaker about 1600. Moray writing to Huygens on September 19, 1664, says that there is in the King's Chamber a spring clock which does the same thing, made by Fromanteel *five or six years previously*, but the small spring which actuates the crown-wheel only rewinds every half-hour. He had also seen a little watch which also does the same as the King's clock. One would have thought that this would be a blow to Huygens, who has hitherto been credited with the first invention of a remontoire ; but in his reply of October 10 he admits that "it is quite true that long ago there were clocks with two springs, the larger periodically rewinding the smaller ; but it is quite a different matter to make them with a *weight* in such a way that, whilst the little weight is being rewound it never ceases to preserve exactly the same force for causing the revolution of the crown-wheel to which it is directly 'hung.'"

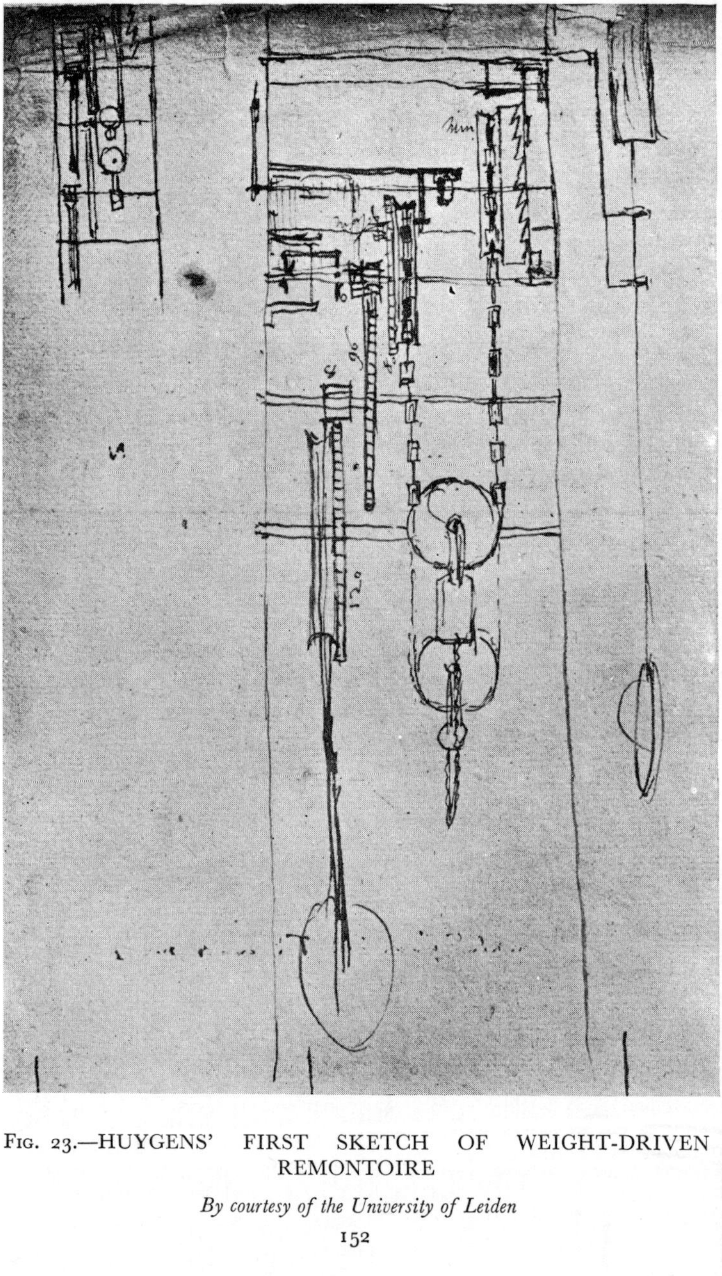

FIG. 23.—HUYGENS' FIRST SKETCH OF WEIGHT-DRIVEN REMONTOIRE

By courtesy of the University of Leiden

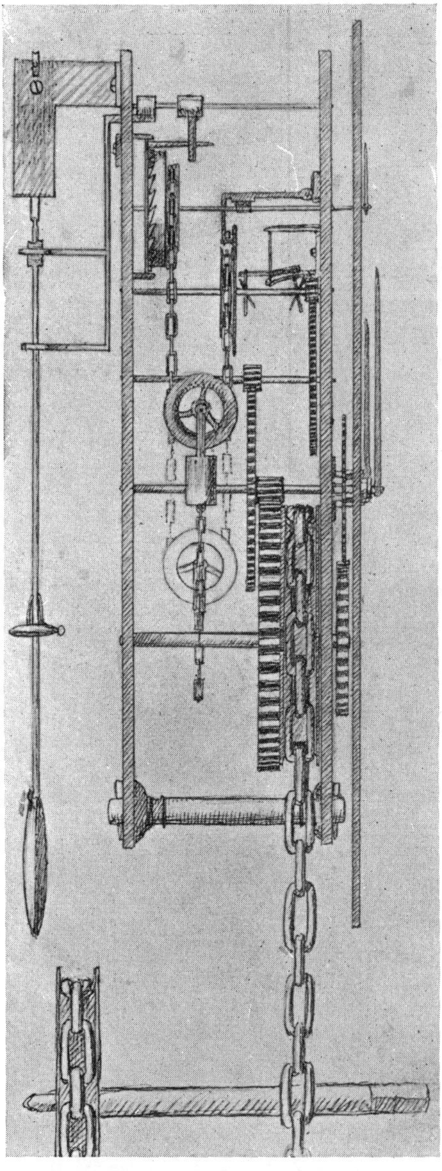

Fig. 24.—Huygens' Final Design of Remontoire

over a new departure in construction, without divulging his idea to Kincardine, and whilst awaiting the result of the trials of the latter's crutch. He must have found difficulty in working out the details, because the first allusion to his "new invention" occurs in a letter from his brother on December 20, 1663, and replying on December 28, Huygens asks that it be kept secret. The new invention is often mentioned in correspondence with Moray in 1664, but without any indication as to its nature. Nevertheless, we find Moray writing already on August 5, suggesting that Kincardine's gimbal arrangement and double crutch will be just as necessary for the new clock; he sees no objection to applying for a patent, and "there will certainly be no dispute as to your respective interests." Thereupon, on August 19, Huygens informs Moray that Severijn Oosterwijk[1] has now completed the clock, and he gives him a general description of the action.

By this time several rough sketches had developed into a beautifully finished drawing, which occurs in MS. Book "C"[2] (Fig. 24).

The length over all of the frame in this design is $4\frac{1}{8}$ inches, the clock, as constructed, being no doubt of considerable size. For the purpose of explaining the mechanism I have made a rough diagram of the upper part of the clock, the portion illustrating the escapement being drawn to scale, whereas the remontoire device has been compressed, as comparison with the original drawing will show.

The action of the escapement, as I understand it, would be as follows:

A is the crown-wheel with its arbor B pivoted in the frame-plates C, C′. D and D′ are the pallets, the verge of which is pivoted in the cocks E, E′. Attached to the upper end of the verge a rod F projects at right angles and passes through an aperture in the piece G. This piece and the piece H are fixed tight on the arbor I, pivoted into the frame-plates C, C′. The piece H is united to and forms one element with the crutch J K, with its two forks L and M. N represents one of the cycloidal cheeks. O is the suspension cord, to which is hooked the pendulum rod P P, with a flat R passing through an aperture in L and through a hole in M (Bruce's invention). When the pendulum is set swinging, the power, derived from the weight suspended below the chain pulley, causes the 'scape-wheel A to revolve towards the spectator of the diagram, the pallets D D′ engaging alternately in the teeth of the wheel. This alternating motion of the verge is communicated by the rod F to the piece G, and consequently to the crutch H J K, and thence through the forks L and M to the pendulum. Obviously, this action

[1] Coster had died towards the end of 1659, and Huygens then entrusted his work to Oosterwijk.

[2] The drawing appears on p. 13, the first dated page, February 22, 1665, being 28. As the clock was completed in August 1664, the drawing must have been made later in a vacant space, unless it represents a further improvement in the design.

would cease when the weight had descended to its full limit as shown in faint outline (Fig. 24), and the original position by rewinding must be restored. This is effected by the force of the heavy weight suspended from the great wheel acting through the

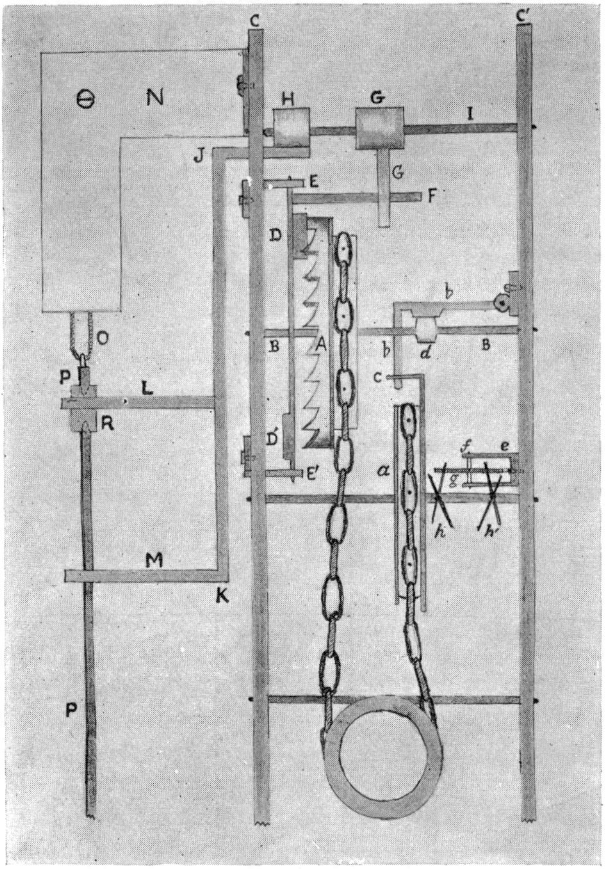

Fig. 25.—Diagram of Remontoire Action.

train of wheels and pinions on the chain-wheel *a*. It will thus be seen that the escapement is kept going evenly by its own little weight, and independently of the main train and weight, and is thus unaffected by irregularities resulting from defects in a long train of wheels.

The clock was calculated by Huygens so that the 'scape-wheel should revolve in half a minute with the pendulum beating half-seconds.

The action of the remontoire mechanism was not so easy to explain, and I have to thank Lieut.-Commander Gould for a lucid interpretation, which fits exactly with Huygens' design, and I have therefore embodied his sketches in my diagram.

A framework *e* carries in its ends a pivoted rod *f*, into which is fixed at right angles another rod *g*. Passing through holes in the arbor of the chain-wheel *a* are a pair of spikes *h h'*, each pair in the form of a cross, and so mounted that, viewed axially, each spike of one set intersects the spaces between the arms of the other set. I now give Lieut.-Commander Gould's explanation :

Let us start the pendulum swinging, the 'scape-wheel turning towards us during its half-minute revolution, and its driving weight descending. Meantime the chain-wheel *a* is locked by the *hinged* arm *b*, its downward projection taking against the pin *c* which forms an integral part of the wheel *a*. On the arbor of the 'scape-wheel A there is a cam *d*, which with the revolution of the arbor gradually lifts the arm *b*. When the 'scape-wheel has completed its revolution since the last unlocking, the cam *d* has lifted the arm *b* high enough for its projection to clear the pin *c*, and the wheel *a* is now free to turn in the opposite sense to the 'scape-wheel, thus, through the force exerted by the main train, rewinding the small driving weight through the distance it had descended. Theoretically this is what would happen ; but so nice an adjustment would not succeed in practice.

In the absence of any other provision the wheel *a* would revolve with rapidly accelerating velocity ; and, after it had made a complete turn, the pin *c* might still find the arm *b* held up by the cam *d*, and the train would continue to run on, throwing the whole action into confusion.

There must be a margin of safety ; the arm *b* must be lifted slightly higher than is necessary to unlock, whilst at the same time the cam *d* must get clear of the arm *b*, allowing it to drop before the pin *c* comes round again.

To ensure that these operations take place without fail it is necessary to give time by checking the too rapid revolution of the wheel *a*. This, in Huygens' design, is accomplished by means of the spiked arbor. When the wheel *a* is released the spikes *h*, *h'* knock the rod *g* backwards and forwards as the arbor revolves, thus setting up a retarding action on the train to such an extent as to make the re-locking of *b-c* safe.

If we refer to Huygens' first rough sketch (Fig. 23) we shall see that an alternative method of controlling the revolution of *a* is indicated in faint outline, consisting of a fly on a separate arbor, geared up from a toothed wheel affixed to the wheel *a*. But the sketch also shows the " idle escapement " of the spikes, to which he finally gave the preference.

There does not appear to be any record at the Patent Office of this new invention, on which Huygens built his hopes, having been patented in England, but such must have been the intention, because the question of Kincardine's share had again arisen. Huygens, writing to Moray on August 8, 1664, says : " I know that there will be no further dispute as to Kincardine's interest and mine now that he has left it to you. But what will he say if the clocks so far employed do not prove satisfactory [this refers to Holmes' second voyage], and that he has to have recourse to *my new ones* ? " Moray, as we have seen, had already given it as his opinion that Bruce's devices would still be necessary ; and on November 7 he informed Huygens that the name of Abraham Hill would be inserted in the patent, and he will execute the relative assignments. For the text of the original patent, dated March 3, 1665, see note on p. 146.

Meantime a request had been preferred to the States-General for an *octrooi* in Holland in the joint names of N. Brus and Christia*e*n Huygens ; but when the *octrooi* was granted on December 15 only Huygens' name appeared. It may be seen in the State Archives ; and, as it states that " a little weight inside the works alone drives the crown-wheel and is rewound from time to time by the great weight," it is plain that a patent for the remontoire clock was in question in England as well as in Holland. The States-General's examining committee probably satisfied themselves that Bruce's share in the invention was confined to the crutch, and therefore declined to accept him as co-patentee.

In his letter to Jean Chapelain of February 5, 1665, already quoted, Huygens adds : " I have advertised this week that the clocks will be offered for sale, with the necessary instructions for their use." [1]

A *privilège* was also obtained in France early in 1665, as to which Huygens, writing to his father on March 5, says that Kincardine must have some share, "*je ne scay pas encore quelle, car la moitiè seroit trop.*"

On May 20 Huygens executed a Power of Attorney at The Hague, empowering Chapelain to sell and assign this French *privilège* on such conditions as his attorney may think fit.[2]

It was in this year that Louis XIV, through Colbert, addressed an advantageous offer to Huygens to take up his residence in Paris, and accordingly he spent the years 1666–76 in that city. As a consequence, succeeding trials were made on French ships ; though in any case the naval warfare which was in progress between the English and Dutch put an end to the employment of English vessels for testing the new invention.

Huygens states ("Horologium Oscillatorium," p. 48) that after Holmes' trial "several experiments were made by Dutch and French ships with varying result, due more often rather to the carelessness of those to whom the clocks were entrusted than to the fault of the clocks themselves."

The first recorded trial appears to have been made by a certain de la Voye, chosen for the purpose by Colbert. This was in 1668, when he gave no satisfaction. On the other hand, he made "very good observations," according to Lodewijk Huygens, when he accompanied the Duc de Beaufort on his expedition in 1669 to expel the Turks from Crete.

On this occasion Huygens relates that weight-driven clocks with a 9-inch pendulum were employed, enclosed in

[1] These instructions, originally in Dutch, are reproduced in Latin as "Brevis Instructio" in "Horologium Oscillatorium," pp. 195 ff., and in English translation in "Philosophical Transactions," vol. 4, No. 47, pp. 937–53.

[2] The official copy of this power may be seen in the Municipal Archives at The Hague, signed "Chr. Hugens de Zulichem."

heavily weighted cases. Observations were made to determine the difference in longitude between Toulon and Crete, and were repeated for the return journey, and found to be approximately the same, " which agreement is the surest sign of accuracy."

According to these trials the difference in longitude was determined as $20\frac{1}{2}$ degrees ; whereas in reality it is only about $18\frac{1}{2}$ degrees, an error of some 120 geographical miles.

The clocks used on this occasion would seem to have been much the same as " Bruce's " ; because Huygens goes on to say in " Horologium Oscillatorium " that it was then that he set about devising the remontoire clock. It is difficult, however, to harmonize this statement with his letter of February 1665, saying that these clocks were now on sale ; all the more so, because Huygens wrote to his brother on December 4, 1667, that " the clocks made in Paris are like ordinary clocks *without the little chains*, as the latter give trouble on the least accident happening." Moreover, we have seen that the drawings appear to date from 1662.

The next recorded trial was made by Jean Richer, the astronomer, on a voyage to Canada in the summer of 1670. The results in no way came up to Huygens' expectations, but he still did not despair of success. He attributed the failure to the observers' " nonchalance," and to Richer's sea-sickness. Moreover, both clocks stopped during a storm. The clocks for this voyage seem to have been specially made, because a letter of Lodewijk Huygens says that Christiaan told him on May 13 that seven changes were made in the marine clocks, amongst which was the removal of the small chains. As far as can be gathered from the correspondence the remontoire, if it ever was tried at sea, proved to be unsuitable for the purpose " *a cause de l'embaras* " ; but it may have been reserved for astronomical observation.

In February 1672, an expedition was dispatched by the

Académie des Sciences to Cayenne, with Richer as observer. On this occasion the object was to make geodetic investigations, and nothing is said about the longitude. Reaching Cayenne in April, Richer remained there until May 1673.

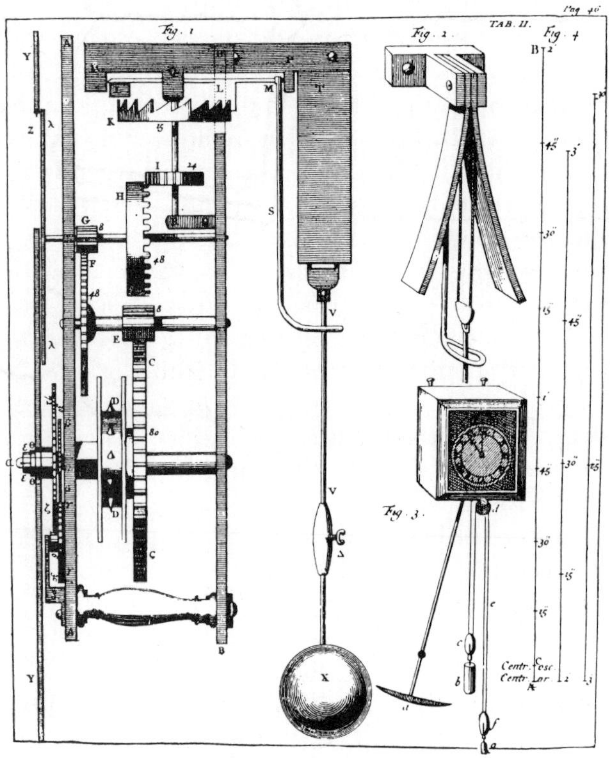

Fig. 26.—Huygens' Standard Clock.
From " Horologium Oscillatorium "

It was on this occasion that he found that his clock, regulated to beat seconds at Paris, lost about $2\frac{1}{2}$ minutes daily.[1]

What clock was used for these observations does not appear ; but as it was one beating seconds, I think that we

[1] This was afterwards confirmed by Varin and Deshayes, and was first explained by Newton in " Principia " as being due to diminution of gravity, owing to protuberance of the earth at the Equator (" Encyclopædia Britannica," 11th edition, vol. 8, p. 802).

shall not be wrong in assuming that it was *the* clock described in "Horologium Oscillatorium," and illustrated in Plate II, i.e. in 1673 [1] (Fig. 26).

Huygens states (p. 36) that this clock was fitted with a 3-foot pendulum. This, he explains, must not be understood to be the ordinary foot, but what he terms "*pes horarius*," which should be used as a standard of measurement for all time.[2]

The crown-wheel is now placed horizontally, the pendulum swings between cycloidal cheeks and the rod passes through a hole in the crutch, and Huygens' maintaining power keeps the clock going during rewinding. Except for the last feature it is virtually Coster's original clock.

In view of the very elaborate explanation that is devoted to this clock in Huygens' *magnum opus*, it must be regarded as his definitive design of a standard clock for use on land.

There is no further record of trials until 1681,[3] when Varin and Deshayes with du Glos were sent out to find the longitude at different places. They went to Cape Verd and St. Thomas, taking with them instructions drawn up by Cassini[4] for managing the clocks and accurately noting time ; and in each place they were to measure the length of the seconds pendulum exactly in order to ascertain how much influence " the earth's rotation " exercised on it.

Similar instructions were also given to Fontenay, who went to China about the same time.

As these observers took clocks with a seconds pendulum we may conclude that they used the standard clock for

[1] A clock made in accordance with Huygens' design (except as to the dial) may be seen working in the Science Museum.

[2] The length of a pendulum beating seconds had been proposed by Wren for this purpose.

[3] This account is given by J.-B. Duhamel (1624–1706), in his "Regiæ Scientiarum Academiæ Historia," 1701. He was Perpetual Secretary of the Académie des Sciences.

[4] Giovanni Domenico Cassini (1625–1712), the astronomer, was appointed by Louis XIV as curator of his new observatory, and was naturalized in 1673. He discovered four satellites of Saturn, and in 1675 the division of Saturn's ring. He published an account of Jean Richer's geodetical observation in Cayenne. Elected F.R.S., May 22, 1672.

observations on land. What they employed for determining the longitude does not appear.

The above records show that during Huygens' residence in France the trials of his inventions were made in French ships, the clocks being made by French makers, notably

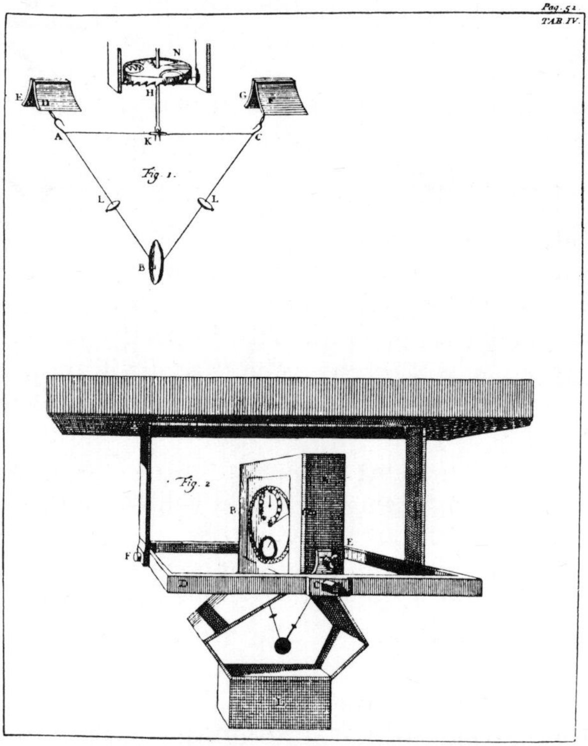

Fig. 27.—Huygens' Triangular Pendulum Clock.

by Thuret; but it would have been strange if his own country, as a great maritime nation, had not sought to profit by his inventions. That such was the case emerges from a letter to Pelisson[1] written as far back as 1679, in which he says:

The last trials with pendulum clocks at sea have not been wholly unsuccessful; but, as they necessarily suffer from the

[1] Paul Pelisson (1624–93) became historiographer to Louis XIV.

motion (*agitation*) of a vessel, there is more hope of success with balances with a spiral spring, but made on a large scale, because accuracy is proportionately increased, and it would be well worth while to make this test (see Fig. 28).

Although no record of a trial at sea of the spiral spring clock of 1675 has so far been found, the possibility need not be dismissed. What seems certain, however, is that the East Indies Company had made use of the long pendulum clock, because Huygens writes to Fullenius on December 12, 1683, that, " at the request of the Company, he had undertaken the construction of clocks to determine the longitude, possessing as constant a regularity as those with the 3-foot pendulum, but such as should not be disturbed by the motion of the sea." And on December 4 he had made entry in the MS. book F of his " Adversaria "[1] : " εὕρηκα. Pendulum cylindricum trichordon. Invented after my discovery that with cold the *spiral spring* accelerates the motion of the clock. Here we have the effect of the spring without the spring."[2]

As usual, the new invention was based on mathematical principles—not so easy to translate into practice. Briefly, the " pendulum " consisted of two disks of equal diameter, the upper one being fixed, and the lower attached to it by three threads (*fila*) equally spaced, so that when freely swinging the lower disk would rise and fall in isochronous rhythm. A ring, wider than the lower disk, was attached to it, which carried small weights which could be shifted for regulating the speed of swing. The length of the threads was to be $9\frac{1}{2}$ or 38 Rhenish inches, *i.e.* $39\frac{1}{8}$ English inches for a seconds beat and one-fourth for half-seconds. The rough sketches and description are not sufficiently precise to indicate in what way the " pendulum " was connected with the going-train. It appears, however, that he delivered

1 The MS. book is preserved at Utrecht.

2 It should here be mentioned that even in 1690 Huygens was not acquainted with the dilatation of matter by heat ; for he then denied that the pendulum rod would expand from the heat of the Equator.

a model to the clockmaker van Ceulen, so that he might alter the clocks that he was making for the East Indies Company. Whether the device was ever used I am unable to say. I have merely introduced this brief account in order once more to demonstrate Huygens' indomitable perseverance in his endeavours to solve the elusive problem of determining the longitude.[1]

In spite of all discouragement, Huygens would not acknowledge defeat. He was at last convinced that a weight-driven clock with a long pendulum was useless for withstanding rough usage at sea, and his next mechanism was constructed on quite different principles, as described in "Horologium Oscillatorium," p. 50, and illustrated in Plate IV (Fig. 27).

Huygens' description may be translated as follows :

> The pendulum has the form of a triangle, to the reversed apex of which is attached a " lens " of lead. The angles of the cord are suspended between cycloidal cheeks. The base of the triangle at its central point receives a forked rod, and is moved by it, the rod itself deriving its motion from a crown-wheel, set horizontally. The train of wheels is driven, not by a weight, but by a steel spring enclosed in a drum. . . . The " little lenses " L L are for regulating the speed of the pendulum.

The method of suspension from below the ship's deck, combined with the 50-lb. weight attached to the bottom of the case, and the thickness of the axles C, C, were to "prevent any excessive freedom of movement in the clock, and to secure that, if by chance it should be disturbed by any serious shock to the ship, it would at once return to rest and to the perpendicular."

The pendulum was only 6 inches in length.

It had been intended to send these new clocks with Richer on the expedition to Cayenne, but owing to Huygens' illness he had not been able to get them ready in time.

[1] I owe the above particulars to Dr. Vollgraff's recent study of the manuscript. We may feel sure that in a later volume of the " Œuvres " he will be able to elucidate what is at present obscure.

So far as is known, fourteen years[1] elapsed before this clock was tried, when it was taken on a voyage to the Cape of Good Hope by Thomas Helder, accompanied by de Graeff, in 1686 on behalf of the East Indies Company. Huygens handed to Helder on April 23 a lengthy instruction as to the use of the clock, and how it should be taken down and put together again, if necessity should arise. From this it would appear that it was fitted with a spring remontoire, discharging and rewinding every thirty seconds, the crown-wheel revolving once in a minute and carrying the seconds hand.

What the result was on the voyage out I do not know ; but, as to the return voyage from the Cape to Texel in 1687, Huygens admits, in a report to the Directors of the Company, that the longitude was not well determined, which he attributes to a systematic error, the rotation of the earth and its influence on the pendulum not having been taken into account ; by correcting this error the result was good, though the two clocks did not always keep time together. Helder and de Graeff also submitted a report.

Another, and, so far as the records show, a final trial was made with these clocks with the triangular pendulum. On this occasion de Graeff left for the Cape on December 28, 1690, taking the same instructions as those given to Helder. A letter from the East Indies Company, dated November 16, 1692, states that " the clocks have not proved such a success as we had hoped for." In his official letter to the Company of March 6, 1693, Huygens—after criticizing de Graeff's journal, finding faults of calculation, and disapproving of his way of using the clocks—does not press for their further employment, because—he concludes—

I have on this occasion invented something quite different and incomparably (*ongelijck*) better, which I have in hand at

[1] As a reason to account for this long interval I offer the suggestion that Huygens was elaborating the spring remontoire, which was finally incorporated in the original mechanism. The fact that in his description Huygens makes no mention of a remontoire seems to point to such an explanation.

the present moment, whereby any little difficulty in the use of this invention will once for all be removed, of which in due course I hope to give Your Excellencies further particulars, and remain, etc.

What this new invention was has not yet been, and perhaps never will be, determined. Was it perchance a perfected spring balance? There is, however, no record of a trial at sea ever having been made.

We have seen that Huygens' original invention of the spiral spring in 1675 did not yield the result that he expected on account of the influence of temperature, but his letter to Pelisson of 1679 shows that he was still hopeful of success. May it not be that he was still working on this problem—the germ of the modern ship's chronometer? The trials to the Cape show that he had at long last abandoned the weight for a spring drive. Why then did he not also abandon the pendulum for the spiral spring balance?

The answer is, maybe, that to discard the pendulum, *his* invention as he held, would be in some sense an acknowledgment of defeat, whilst at the same time casting a shadow over the lustre with which the publication of his world-famous "Horologium Oscillatorium" had invested him. It is remarkable that he adheres to the crown-wheel and ignores the anchor escapement up to the last.

Such frequent references to this work have been made in these pages that Dr. Hooke's opinion both as to the application of the pendulum and the endeavour to determine the longitude at sea must be of particular interest.

In the "Philosophical Transactions," No. 95, of June 23, 1673, there appears "An accompt of Christiani Hugenii 'Horologium Oscillatorium'"; but this review contains no criticism of Huygens' claims or statements. Probably the "accompt" was written by Huygens' friend, Oldenburg.[1]

[1] Heinrich Oldenburg (1626–78) was a native of Bremen. He first visited England in 1640, and, after travelling abroad, returned as diplomatic agent of Bremen, finally settling in England. He became the first Secretary of the Royal Society.

On examining the manuscript remains of Dr. Hooke at the British Museum I had the peculiar good fortune to discover his unpublished Notes for a lecture to the Royal Society on the subject, provoked no doubt by the colourless account above mentioned.[1] In view of the bitter attack upon a fellow-member of great distinction, it is difficult to suppose that it was ever delivered ; but its very frankness makes it all the more valuable as a contemporary contribution to the controversy on the part of the man who, of all others, from his intimate acquaintance with the subject, both theoretically and practically, was most competent to offer destructive criticism.

The following is a transcription of the Notes *verbatim et literatim*, omitting only a complimentary exordium address to the Royal Society :

I have lately Received from the Inquisitive Hugenius van Zulichem a book written by himself containing a Description of severall mechanicall & mathematicall Inventions Intituled Christiani Hugenii Zulichemii Const. f.[2] Horologium Oscillatorium sive de motu pendulorum ad Horologia aptato demonstrationes geometricæ.

There are in it indeed many things very ingenious and very usefull but there are not wanting also severall things that are of a quite contrary nature as I shall shew you by some few observations which I have made in the cursory reading of it having not yet had time to examine every particular thereof more strictly.

And in the first Place the Authour gives us an account how about 15 years since he first published his Invention of applying a pendulum to a clock and thinks thereby he hath sufficiently secured and warranted himself to have been the first inventor thereof because there was noe body before him that had made publication thereof to the world, and is very unwilling to Allow Gallileo any share in the honour of the Invention. Whether Galileo or his Son did find out away of applying it to a clock I can not affirme but sure I am that the Greatest excellency of the Invention is to be ascribed to Galileo who first found out that the Vibrations of a pendulum were very

[1] British Museum MSS. (Sloane 1039, folio 129).

[2] *I.e.* Constantini filii.

neer of æqual Duration. Nor Is mersennus or Ricciolus to be deprived of their share in further examining & promoting the Doctrine of pendulous motions nor that French authour who writt animadversions upon Galileos mechaniques, who does not only speak of the applycation of a pendulum to clocks, but also enigmatically Describes a way of using it at Sea for the finding out the Longitudes, nor Indeed after the knowledge of the æquality of pendulous motion was it Difficult to find out away of applying it to Clocks. Dr. Wren, Mr. Rook, Mr. Ball & others made use of an Invention of Dr. Wrens for numbring the vibrations of a pendulum a good while before Mon^r. Zulichem publisht his and yet did not cry εὕρεκα, and, I my self had an other way of continuing and æqualling the vibrations of a pendulum by clock work long before I heard of Mr. Zulichem. And yet though I think I doe deservedly much preferr it before M^r. Zulichems way, nay though æqualed wth a Cycloeid yet I have not either cryd εὕρεκα or publisht it and yet I think I can produce a sufficient number of credible witnesses that can testify for it about these 12 years. Soe that the argument that he soe much Relys upon to secure to him the Invention is not of soe great force as to perswaid all the world that he was the first & sole inventor of that first particular of applying a pendulum to a clock.

The next thing wch he mentions is his invention that all the vibrations of a pendulum moved in a cycloeid are of æquall Duration. This for ought I know he is the first Inventor of for I never heard of any one that claimed the honour from him of it. It is indeed an Invention very extraordinary and truly excellent and had been honour enough for him Justly to have gloried in the happinesse thereof, and I believe there is none that would have gone to have deprived him of his Due praise, But he should also have Remembred that Golden Rule to doe to others as he would have others doe to him & not to have Vain-gloriously & most Disingenuously Indeavoured to Deprive others of their Inventions that he might magnify himself, and wth the Jack Daw pride himself in the plumes of others, which how much and often he hath Done in the rest of the book I shall Indeavour to explaine.

But before I come to these particulars which are indeed noe ways pleasant to me were there not a necessity and duty incumbent on me to Doe it Give me Leave a little to animadvert upon those two Inventions wch for ought I know may Justly be his own. that is the way of applying a pendulum to a clock and the æquation of the motion of a pendulum by a cycloeid. For the first I say the Invention is very simple and plaine and therefore soe much the more to be preferred, but yet if thereby the pendulum becomes affected wth every inæquality that the Clock work which it really doth is subject to, and that inæquality be not removed by the æquation of the cycloeid as certainly & experimentally it is not, then be the Geometricall subtilty and demonstration thereof never soe excellent yet it is in it self but a lame invention, and he hath come short by a point if he hath made it dulce & not utile.

First therefore I say his way of applying a clock work to ye pendulum is imperfect for the pendulum is sensible of every inæquality of the motion thereof, and though by reason of ye small proportion of the strength of ye watch work[1] to the strength of ye pendulum it is not sensible in any single vibration, yet in a Longer Duration of time it becomes most sensible when even the best & most accurate pendulum clocke will sometimes goe faster and sometimes slower, sometimes make greater & sometimes lesser vibrations and having not produced any invention to make all the vibrations of æquall extent he hath not yet produced the invention which can be calld the perfection of timekeeping. For in the 2^d^ place his invention of æqualing the vibrations of a pendulum by a cycloeid will noe ways doe it. first because his very first Supposition or Demonstration of the nature of the pendulum in a cycloeid does destroy his Imagined Theory for if the motion of the pendulum untoucht will make æquall vibrations then certainly those vibrations must be unæquall if they are sometimes more and sometimes lesse promoted & sometimes more sometimes lesse hindred by the impelling power of the clock work. for though the vibrations of a pendulum depending of itself by its own proper gravity will passe either a greater or a lesser arch in the same time the swiftnesse in the greater being to the swiftness in the lesse as the length of the greater to the length of ye lesse but if that swiftnesse be altered by an other impelling force then that of gravity then that vibration that hath most impelling force will passe the vibration in shorter time & that wch hath lesse impelling force will passe it more slowly. and in not only in the Descending part of the cycloeid but also in the ascending. for though the pendulum left free will ascend either a larger or a shorter arch of the cycloeid in the same time according to the degree of velocity it hath in the perpendicular. yet if that free motion be stopd by a stronger or weaker check of the clock work it shall much sooner or later make its returne and consequently all that æquality demonstrated in the theory of the Cycloeid,

[1] In those days the going train of a clock was called the "watch." (See Derham, *passim*.)

is by the ill application of the strength of ye clock work to continue that motion Destroyed. soe that we are not anything further promoted towards æquality by this new propriety of the Cyclœid. for I judge a pendulum moving in a circle to be a much better measure of time. My reasons are first because as a greater impuls from the clockwork does make a larger & quicker vibration soe on the other side is the larger vibration in its own nature more slow and consequently doth æquale and adjust the greater velocity and swiftnesse given by the Impulse of the Clockworke. but because this doth not exactly Adjusten and æquall the motion of the pendulum it being observed by experiment that a plain pendulum clock will be made to goe slower by hanging more weight to move ye watch work and to goe somew[t] faster by taking away somew[t] of that weight for that experiment does seem to hint some middle way of suspending a pendulum between that of a common pendulum & that of a cycloeidicall pendulum. which how it may done I shall hereafter shew. But in the next place my second Reason why I preferr a common pendulum beyond that of Mon[r] Zulichem is because that moves only upon one center namely the center of suspension, the rod of the Pendulum being stiff and not at all plyable between the points of Suspension and ye Ball, and consequently can have but one kind of Inæquality, whereas that of Mon[r] Zulichem being suspended partly by threads ribands or pliable materiall in order to the bending about the Cycloeid cheeks and partly also by a stiff Rodd or plate is subject to an other great inæquality namely to a bending at the place where the stiff & plyable parts are Joynd together. and this is not notionall only but very visible to the eye especially if the check be great that is given it by the watch part, soe that after all this paines for the adjusting the pendulum after Mon[r] Zulichem's way we come short of that Idea of Perfection in the measure of time which his Geometricall Demonstrations would Insinuate nor is possible by any of the ways he hath mentiond or Insinuated to make an exact instrument for that end we must therefore seek somew[t] Further if we designe to doe any thing accurately in Astronomy not now to mention Geography & the Longitude because I shall come to that by and by when I come to consider his method of carrying the pendulum clocks by Sea. which is the next he proceeds to Discourse of after the Description of his pendulum clock in Generall. And herein he deals very unjustly wth that noble Person the Earl of Kingkarden who was the first man that attempted to carry the pendulum upon the Sea and to Apply it for the use of Navigation and finding the Longitude describing the story thereof in such a manner as if he himself were the inventor of it whereas tis certaine that he himself drempt of noe such thing till my Lord Kingkardine had made tryall of it, and upon the early giving him an account of y[e] Successe he was much surprised at the novelty thereof as not thinking it possible as I was heretofore acquainted by the earl himself. when he first brought it into England. nor was that Instrument made ad nostrarũ exemplum as Mon[r] Hyginus would insinuate but of a particular forme of y[e] Earles owne invention much like this here Described to witt in the forme of a very thick Quadrant Loaded at the Limb wth a great weight of Lead & suspended at the right angle by a ball which moved in a Socket which was fastend the underside of y[e] Deck of y[e] Ship. this Instrument being made in holland by y[e] earle was brought over wth him into England in order to make tryall of y[e] Experiment but the vessel being but small and y[e] weather pretty rough and stormy the watch after some short time stood still. soe that nothing could be certainly concluded from it. In February or March 1664 as I remember my Lord Kingkarden having gotten another made here in England did together with my Ld. Brounker Sr. Ro Moray & myself make a further tryall of them wth some of y[e] Kings Pleasure boats but not wth soe good successe as was expected. one of y[e] watches by the shaking of the boat in the carriage from white hall to greenwich ceasing to goe, and the watches afterwards not keeping exact proportion to one another. they were afterward sent by Sr. Robert Holmes to Guinny. and an account returnd thereof somew[t] Like that printed by Hugenius made by one of the Captaines giving indeed a very favourable account of their performance but concealing all their faileurs and miscarriages whereas another person that was in the same ship gave a relation very differring. which relation was concealed and the other printed. all this was done before Mon[r] hygenius had contributed anything towards this use of them & you may find by his Letter w[ch] was printed in y[e] s[d] transaction that Major holmes his Relation of y[e] Pendulums surpassed his Expectation, and that he did not imagine that the watches of this first structure would succeed soe well.

All that I find he added to the Invention was the putting a chaine wth a weight over the ballance wheel which was wound up by the Clock work which is indeed not soe good as that made use of by my Lord Kingkarden as I could sufficiently both by Reason and Experience Demonstrate were it necessary but I think it wholly needlesse as being sufficiently satisfyed that neither the one way nor the other will be of any great use at sea much lesse certainly to find the Longitude nor will the third that he hath added doe it that having some inconveniences greater than either of the other and noe way removing any of the former. as I shall hereafter shew. however it seems he soe farr prevailed in Holland that the States did Receive his propositions when he Desired of them a patent for those new watches and the recompense set apart from the invention in case of successe and that wthout any difficulty they had granted his Request commanding him to bring one of those watches into their assembly and to explicate to them his invention and application thereof to the Longitudes which (in the foremen-tioned epistle printed in the first Philosophical transactions) he says he had done to their Contentment. But notwthstanding we doe not find by this last book of his that he hath yet received that reward though it be now some nine years since. nor will he ever justly receive it unless he find out some other way than what he hath yet mentiond.

It was in ye year 1658 & 59 that I indeavourd to Obteine a Reward from the States for Discovering A practicall and certaine way of Discovering the Longitude of Places by Sea and Land but upon inquiry concerning such a reward I was answerd that there was never any such thing, and that it was nothing but a groundlesse fiction the like answer I received concerning the Imaginary præmiums in Spain & france. nor could I find any better incouragement here in England when I propounded it in the year 1660. though yet I am sure my way was certaine and very easily practicable and might possibly be the occasion of putting my Lord Kingkarden in mind of making tryall wth the pendulum Clocks. But this I mention only by the by and shall hereafter speak more fully thereof when I shall publish that invention and the history thereof. But to proceed that experi-ment that he relates to have made wth two pendulum clocks made wth chaines about the ballance wheels is very strange, and indeed as great an argument against their æquall motion when carried in a ship as he could possibly have brought his words are [*quotation omitted by Hooke*] for since tis evident by his owne observations and experiments that soe very little a motion as was communicated to the beame by wch ye clocks were suspended by, one of the clocks had soe great an influence upon the other clock as to make it in spite of ye unæquall length of ye pendulums to vibrate together what change must not soe great & irregular a motion as that of a ship in fowle weather work upon them. nor will his new sort of pendulum suspended by threads triangular wise or the suspension of the Clock upon the same center wth that of ye motion of the Pendulum at all prevaile in the case. for that sort of pendulum being held by those threads stiff in that posture if by the stiffness of the center pinns F & G the clock be any way out of Perpendicularity the pendulum becomes an inclind pendulum and consequently moves soe much ye slower and consequently receives a variation from the motion wch is the thing which he indeavours to obviate but hath noe ways found a means to doe.

And indeed if one well considers the manner of suspension it seems very rationall to suppose this manner of suspension much worse than my Lord Kingkardines, for in this of Hugenius the vibration of the machine being upon the same center wth ye vibration of ye pendulum the motion of the one is the more likely to confound the motion of ye other, then when the motion of the machine is upon a point a good way removed from the center of ye vibration of ye pendulum either above or below. And this much may suffice for the things mentiond in the first part of his book wch is a Description of the Instruments themselves.

[*End of Manuscript.*]

Hooke must have addressed a milder criticism to Sir Constantijn Huygens, who forwarded it to his son ; for an English translation of the latter's reply exists in the Letter Book of the Royal Society in the following terms (August 17, 1673) :

I am obliged to the civilitie of Mr. Hooke for what he writes to you concerning my Book. . . . I do not wonder at what he sayth to have observed touching the insufficience of the Pendolo's, to find the longitude, because he hath only seen the experience of those which the Earl of Kinkardin had caused to be made and such like, for they had yet very considerable defaults. The last form I have reported in my Book is a great deale better and I am still in good hope of it, expecting that a tryal be made of it at sea, which without these unhappy warres had been done by this time. . . . I beseech you to communicate all this to Mr. Hooke.

Let me conclude with a passage from the short " Life of Huygens," in Latin, prefixed to 's Gravensande's edition of the " Opera Varia " of 1724 :

He was persuaded that his clocks could be used at sea, and that nothing more was requisite on board for determining longitudes. For it is well known that the solution of this useful problem of the longitude—a need long felt, and maybe destined still to be—depends on the exact measurement of time.

It was not, however, sufficient to determine the motions of clocks by fixed laws ; but, as he himself admitted—to preserve an equal motion when the ship was labouring in a storm—*hic opus, hic labor erat*. But he always hoped to be able to overcome the difficulty ; many were the trials he made ; and, in order to reach the goal he persevered with fresh attempts almost to the day of his death. Nevertheless, albeit he was unable to attain it, whosoever peruses the first volume of these works may judge with what genius, with what penetration he pursued the subject. Not all his attempts have, however, been made public.

In this quest of over thirty years his pertinacity in the face of continual disappointment left him sanguine to the last, and calls for our profoundest admiration. His hopes of success were shattered, but to him—"*summus Hugenius*," as Newton called him—the lines of Horace might well have been applied :

Justum et tenacem propositi virum
Impavidum ferient ruinæ.

CHAPTER X

THE BALANCE SPRING

ALTHOUGH I have said that Huygens' proposed marine clock controlled by a spiral spring led the way to the gradual perfection of the ship's chronometer, it must not be assumed that he was the first to apply such a spring to watchwork. The controversy on this subject has been as keen as that with regard to the first application of the pendulum to clockwork.

Reid,[1] whilst himself giving the credit to Dr. Hooke, quotes the great French mathematician, Jean Etienne Montucla (1725–99), as being the only foreigner who shared this opinion, and he gives a long extract from Montucla's "Histoire des Mathématiques," vol. ii, p. 488, in support of it.

Some confusion arises from the fact that the first watches in which a spring was applied to the balance were called "pendulum watches," the spring being intended to impart isochronous motion to the balance, as did the pendulum to the clock. Moreover, as I have mentioned, the going train of a clock was termed the "watch." Thus Evelyn, recording a meeting of the Royal Society on August 29, 1666, states that "Mr. Hooke produced his new pendulum."

This entry has by some been held, as I think, to refer to the invention of the anchor escapement and "Royal" pendulum; whereas the Minute in the Society's Journais states that "Mr. Hooke produced a new piece of watchwork of his contrivance serving to measure time exactly both by sea and land; of which he was ordered to bring in a description." I think that this mechanism was a *watch* regulated by a spring; because Hooke had already been much occupied with this problem, as we shall see; but whether it was a spiral spring is not apparent.

[1] Thomas Reid, "Treatise on Clock and Watchmaking," Edinburgh, 1826, p. 264.

A further difficulty is caused by the interchangeable use in the French correspondence with Huygens of *montre* and *horloge*.

Huygens, then resident in Paris and enjoying a pension from Louis XIV, applied to the Minister Colbert on February 5, 1675, for a "*privilège*" for his spiral spring balance, and this was granted ten days later. But the Abbé de Hautefeuille at once entered a protest, alleging that it infringed his invention, which in fact was couched in the most general and vaguest terms, with no specific mention of a *spiral* spring. Père Allexandre,[1] a contemporary and admirer of the Abbé, relates that on July 7, 1674, de Hautefeuille "submitted to the Academy of Science a method for regulating the motion of the balance of watches by means of a little *straight* steel spring fixed at one end to the edge of the plate, the other end being so inserted into the edge of the balance as to have freedom for moving to and fro, and thus forming a sort of pendulum." But Allexandre frankly admits that "Mr. Huygens has perfected this invention by making the spring in the form of a spiral."

I have alluded, in the early part of Chapter VI, to de Hautefeuille's diatribe against Huygens, but I think that Allexandre's testimony is sufficient to dismiss Hautefeuille from the controversy. His opposition was, however, sufficient to deter Huygens from further action.

Huygens, writing to a lady on May 6, 1675, says that "de Hautefeuille's invention is nothing but a straight steel spring attached to a clock instead of the pendulum, and set in motion like the pendulum by means of a fork. Little attention was paid to the document at our meeting (i.e. of the Académie), not only because the uselessness of the invention appeared from the author's own admission, but because it was known that others had tried the same thing without success." He then adds that by comparison of

[1] "Traité général des horloges," Paris, 1734, pp. 24 and 243.

Hautefeuille's construction with his own " it will be seen how much they differ ; since, beside a totally different form and application of the spring, I employ a balance that turns on its pivots, and my invention consists in the combination of these two things "[1] (" Œuvres," vol. vii, p. 457). (Fig. 28.)

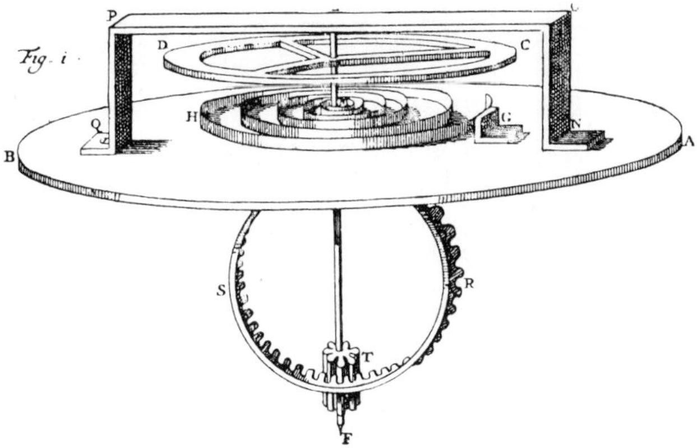

Fig. 28.—Huygens' Balance Spring.

On leaving Paris[2] to return to Holland on July 1, 1676, after five months' illness, he wrote :

Je laissay libertè entiere a tous les horlogeurs de travailler a cette Invention, voyant que le Privilege m'aurait encore coustè des sollicitations pour le faire enregistrer au Parlement, et que mesme apres j'aurois tousjours des proces et des embaras nouveaux.

The preliminary grant was made, as I have said, on February 15, 1675, and five days later Huygens, writing in French as usual to Oldenburg, the Secretary of the Royal Society, offered to *him* or to the Society all his interest in the invention, if it was thought that a patent in England would

[1] The engraving is part of Plate XXIX in the " Opera Varia," 1724.

[2] He remained once more in Paris between 1678 and 1681, but did not again return.

12

have any value. I shall return to this later ; but I may mention here *en passant* that none of Huygens' horological inventions appears to have brought him any pecuniary advantage.

In what then did Huygens' invention consist ? As was the custom of those days, a secret was concealed in an anagram, the solution of which in this case he revealed to Oldenburg in a letter, dated January 30, 1675. It was as follows :

Axis circuli mobilis affixus in centro volutæ ferreæ, that is, the balance staff is affixed in the centre of a steel spring.[1]

The device was described and illustrated in the *Journal des Sçavans* of February 25, 1675. I cannot give a better explanation of it than that contained in Commander Gould's monumental treatise on the Marine Chronometer.[2] He says :

> Huygens' method of applying the spring was original, for he geared up the balance, so that instead of describing, like Hooke's, an arc of 120° or so, it revolved several turns at each beat. This device is termed a " pirouette." It is theoretically objectionable on account of the friction in the gearing, and never came into general use. He found himself, however, baffled by the effect of temperature on the strength of the spring.[3]

It may be added that the delicate task of producing a spring, whether flat or cylindrical and homogeneous throughout, resistant to change of temperature and of correct length to give isochronous motion to the balance in all positions, etc., baffled many a great inventor for long years to come.

[1] The anagram was as follows :

4	1	3	5	3	7	3	1	2	3	4	3	2	4	2
a	b	c	e	f	i	l	m	n	o	r	s	t	u	x

[2] " The Marine Chronometer, its History and Development," by Lieut.-Commander Rupert T. Gould. London, 1923.

[3] On p. 179 of his MS. book F of " Adversaria," under date December 4, 1683, Huygens writes : εὕρηκα. Pendulum cylindricum trichordon : inventum postquam elaterem spiralem frigore accelerare motum horologii repperi. (See also note to p. 163.)

Before proceeding to consider the rival claims of Huygens and Hooke to the invention of the *spiral* spring, it will be of interest to relate that, beside Hautefeuille, other inventors were in the field, a fact which I have not seen mentioned in any work on horology.

On July 22, 1665, Sir Robert Moray had written (in French) to Huygens concerning a matter " which I have never mentioned to you before," namely, a new invention of Hooke—" or rather twenty "—for applying a spring to the balance. Replying on September 18, Huygens says :

I must tell you that, being in Paris in 1660, the Duc de Roannais spoke to me of the same thing, and even took me to the clockmaker (Martinot) to whom he and Monsieur Pascal had communicated this invention, but under oath and promise before a Notary neither to reveal it nor appropriate it ; but I did not consider their mode of application by any means good, and I already (*des lors*) knew of some much better methods. But, apart from not being as easy to put into practice, as is the case with pendulum clocks, I do not anticipate as much accuracy as that which I find in the latter, seeing that the motion of the ship must cause little irregularities in the movement of the spring, which it would be difficult to remedy ; and it is not yet known whether change from heat to cold would make no alteration in the vibrations " (" Œuvres," vol. v, p. 486).

In Huygens' unpublished Journal under date November 11, 1660, he says : " Martinot came to see me and spoke of the invention of the spring in place of a pendulum."

On December 24, 1665, Huygens writes to Moray : " I do not think that the Duc de Roannais thinks any more of putting this invention into practice " (" Œuvres," vol. v, p. 549). What share Pascal had in the Duke's invention does not appear.

Finally, on February 21, 1675, Huygens writes in his " Adversaria," Book V (" Œuvres," vol. vii, pp. 408 ff.) : " I showed my model to the Duc de Roannais, who spoke to me of the invention of d'Alesme, and said that it differed considerably from mine, and was not half as good or as simple." The nature of André d'Alesme's device is not known. He was an engineer.

The next claimant appears in the person of d'Esson, Seigneur d'Aigmont, an engineer, mathematician and engraver. He is known under many variants of his name, as de Lisson, Duson, Desson, van Son and du Son.

Under the last appellation Oldenburg writes (in French) to Huygens on October 7, 1665 (" Œuvres," vol. v, p. 501):

> Monsieur du Son came to see me to-day, speaking of 4 watches that he is having made for the Court, which he says go as accurately as the sun itself; but he was still keeping their secret. However, when asked if it is not by the application of a spring to the arbor of the balance he does not contradict. I think that it is the same thing in fact . . . as that for which Mr. Hook alleges that he knows so many different methods, and as that which was communicated to you some years ago in Paris.

Writing again on January 15, 1666 (" Œuvres," vol. vi, p. 7), Oldenburg says that du Son's method consists of " un ressort bien deliè, pour regler au lieu de Pendule, le Balancier." It must be confessed that this description is very vague. Moray had previously informed Huygens that he would know more about it when the completed watches were sent to him. But there the matter seems to have ended.

As far as I am aware every English writer has attributed to Hooke the honour of inventing the spiral spring. Apart from Hooke's own statements, most authors have derived their information from Derham's " Artificial Clockmaker," first published in 1696, in Hooke's lifetime, under the initials *W. D.* M.A.

The Rev. William Derham, D.D., was a fellow member of the Royal Society, and after Hooke's death he edited some of his unpublished manuscripts. His account, therefore, " Of the Invention of those Pocket-Watches, commonly called Pendulum Watches " was, we may suppose, inspired by Hooke himself, and as such may not be altogether impartial. The chapter (viii) is of some length; but, as a contemporary statement of Hooke's case, it seems desirable to reproduce the salient points in Derham's own language:

The first *Inventer* was that ingenious and learned member of our *Royal-Society*, Dr. *Hook*: who contrived various ways of Regulation. One way was with a load-stone[1]: another was with a tender strait Spring, one end whereof played backward, and forward, with the ballance: so that the ballance was to this Spring as the bob of a Pendulum, and the little Spring, as the Rod thereof. And several other contrivances he had besides of this nature.

But the Invention which best answered expectation, was at first, with two ballances: of which I have seen two sorts. . . . One way was without Spiral Springs, the other with . . . And the excellency of the latter is, that no jirk, or the most confused shake, can in the least alter its Vibrations. [This, he says, is not the case with a single balance.] But notwithstanding this inconvenience, yet the Watch with one ballance and one Spring (which was also Dr. *Hooks* Invention) prevailed, and grew common, being now the universal Mode: but of the other very few were ever made. The reason hereof, I judge, was the great trouble and vast niceness required in it.

[I omit two pages describing the two systems. Suffice it to say that in each the balances had toothed rims, playing in one another. In the better watch each balance had a *spiral* spring.] But "only one Ballance had the Pallets, as the common Ballances have: and the Crown-wheel operated upon it, according to the usual way." [So that the second balance was only introduced to steady the action of the main escapement.]

The time of these Inventions was about the year 1658, as appears (among other evidence) from this inscription, upon one of the aforesaid double Ballance-Watches, presented to K. *Charles* II, *viz. Robert Hook inven.* 1658. *T. Tompion fecit* 1675. This Watch was wonderfully approved by the King; and so the Invention grew into reputation, and was much talked of at home, and abroad. Particularly its fame flew into *France*, from whence the Dauphine sent for two; which that eminent Artist Mr. *Tompion* made for him.

[Derham then refers to the disagreement about a patent in

[1] Wood, "Curiosities of Clocks and Watches," 1866, p. 103, quoting Weld's "History of the Royal Society," says that a loadstone clock and Hooke's magnetic watches were always exhibited as great curiosities on the occasion of the visit of illustrious strangers. This is often mentioned in the minutes of the Royal Society for 1669.

1660 " with some Noble Persons who were concerned for the procuring it."] The same ingenious Dr. had also a Grant for a Patent for this last way of Spring Watches in the year 1675; but he omitted the taking it out, as thinking it not worth the while.

After the Inventions of Dr. *Hook* . . . Mr. *Hugen's* Watch with a Spiral Spring came abroad, and made a great noise in England, as if the Longitude could now be found. . . .

This Watch of Mr. *Zulichem's* agreed with Dr. *Hook's* in the application of the Spring to the ballance : only Mr. *Zulichem's* had a longer Spiral Spring, and the Pulses and Beats were much slower. That wherein it differs, is : 1. The Verge hath a Pinion instead of Pallets ; and a Contrate-wheel runs therein, and drives it round, more than one turn. 2. The Pallets are on the Arbor of this Contrate-wheel. 3. Then followeth the Crown-wheel, *&c.* 4. The ballance, instead of turning scarce quite round (as Dr. *Hook's*) doth turn several rounds every vibration. . . .

Whether or no that ingenious person doth owe any thing herein to our ingenious Dr. *Hook*, it is however a very pretty, and ingenious contrivance ; but subject to some defects : *viz.* When it standeth still, it will not vibrate, until it is set on vibrating : which, tho it be no defect in a Pendulum Clock, may be one in a Pocket-Watch, which is exposed to continual jogs. Also, it doth somewhat vary in its Vibrations, making sometimes longer, sometimes shorter turns, and so some slower some quicker vibrations.

As to Derham's technical capability to criticize the respective mechanisms there can be no doubt ; but it must be borne in mind that he was only eighteen when, in 1675, Tompion made Hooke's watch, and when (as he admits) Huygens' spiral spring movement created a sensation in England. His testimony that the Tompion double balance watch was regulated by spiral springs can hardly be questioned ; but much stronger evidence is needed than the engraving in the watch to prove that the invention was made in 1658.

Huygens' letter of February 20, 1675, offering to present his invention to Oldenburg or to the Royal Society (" Œuvres," vol. vii, p. 423), was read to the Society on the 28th ; whereupon Hooke said that " divers years ago he had had such an invention, and that actually watches had been made according to the same, for which he appealed to the Journal Books . . . and to several members of it."[1]

His indignation at being robbed of the credit of first invention knew no bounds, and, as usual with him, manifested itself in violent language and direct accusation against Oldenburg of having revealed his secret to Huygens. A printed invective next took the form of a "Postscript" to his "A Description of Helioscopes," published in 1675, wherein he alleges that he perfected his invention about 1658, and in 1660 he "was in treaty with certain persons of honour for the discovery thereof upon proposed Articles of Encouragement." (In another letter he gives the names: Lord Brouncker, Boyle and Moray.) He then relates that—

the negotiations were broken off on account of a clause which stated that: If after I had discovered my Invention about the finding of the Longitude by watches, or otherwise (though in themselves sufficient) They, or any other Person should find a way of improving my Principles, he or they should have the benefit thereof, during the term of the Patent, and not I . . . Upon this the treaty was broken off, and I concealed the farther discovery of the other more considerable parts of my Inventions. . . . Indeed Mr. Hugens hath made use of that part I discovered (*i.e.* revealed).[2]

It cannot be gainsaid that Hooke rightly refused to subscribe to such an iniquitous condition; nor can one refrain from suspecting that these "persons of honour" had Huygens in mind as the "other person" who might find a way of improvement.

After referring to his Cutlerian Lectures in 1664 as to "the application of Springs to the Ballance of a Watch" (he said he knew twenty different ways), "and the production of several models at Meetings of the Royal Society," he adds: "Of all these things the Publisher of the Transac-

[1] Birch's "History of the Royal Society."

[2] It was in this "Postscript" that Hooke printed his famous anagram of "Ut tensio, sic vis."

tions (*i.e.* Oldenburg) was not ignorant, and I doubt not but Mr. Hugens hath had an account."

Oldenburg cannot have had a perfectly clear conscience, as the following letter of October 11, 1675, to Huygens (in French) suggests:

Since my last an incident has occurred which obliges me to write this letter. Mr. Hook, having learnt that you had given me permission to avail myself of any interest you might claim from a patent[1] in this country for your watch, has been so bold and impudent as to say publicly that you gave me this permission as a reward for having revealed to you his invention, adding with the utmost effrontery that I am your spy here for communicating to you what ever of moment is discovered here, and that I wanted to rob him of the profit of his invention. . . . Such accusations are as atrocious as false. . . . I have never communicated to you anything about this invention, nor of any other until after you had revealed your own to us. Sir, I beg you to tell the whole truth as fully as you possibly can . . . Sir, after having made use of this letter, put it in the fire, I beg of you.

Huygens—*who kept the letter*—responded to Oldenburg's appeal in a letter to Lord Brouncker on October 31 (" Œuvres," vol. viii, p. 529):

I take the liberty of writing you these lines, being compelled thereto by the interest that I feel in Mr. Oldenburg's honour and my own. . . . I had invented a new clock [*horloge*] of which I sent him the secret. . . . Mr. Hooke said that he had invented a similar thing, as he thought, some years ago. . . . It is his customary vanity to claim the invention of everything. . . . As to the manner of his construction I am still in ignorance. It is therefore false that, as a reward for having revealed something to me, I wanted to confer a favour on [*gratifier*] Mr. Oldenburg; and my sole motive in this is that I thought that he certainly deserved to derive some profit from new inventions which he has taken so much trouble in publishing in his Transactions to the great benefit of science, and moreover because he might have more need of it than any other of my friends in England. . . . But no one will be so unjust as to condemn

[1] A draft of the patent was found amongst Oldenburg's papers, but it was not taken out. It is preserved in the archives of the Royal Society.

him on the simple accusation of such a man as Mr. Hooke who is rather in the present case impelled by his own interest.[1]

It appears that "Helioscopes" was printed without the knowledge of the President, Lord Brouncker, who replies to Huygens on November 8, 1675: "I was confident of ye innocency of Mr. Oldenburg in all that Mr. Hook thought fit to charge him with" and he adds that the "Postscript" "had been corrected and the severe passages expunged. Mr. Oldenburg has done himself some justice in the recently published Transactions."

This defence was officially published in a review of Hooke's "Helioscopes" on October 25, 1675, in which Oldenburg wrote:

'Tis certain that the Describer of the Helioscope some years ago caused to be actually made some watches of this kind, yet without *publishing to the world a Description of it in Print*: but it is as certain that none of those watches succeeded.

Hooke, however, was not to be silenced. In 1676 he published a book entitled "Lampas, or Description of some Mechanical improvements of Lamps and Waterpoises," and to this he attached another Postscript. In this, speaking of Oldenburg, he says:

The publisher of Transactions indeavours to cover former injuries done me by accumulating new ones. . . . How could he be sure . . . *whom I have not acquainted* with my Inventions, since I looked on him as one that made a trade of Intelligence.[2]

It may be asked, if Hooke had never acquainted Oldenburg with his inventions, how he could have betrayed them

[1] Already on June 7, 1675, Oldenburg had requested Huygens to "write three words to Lord Brouncker, representing to him that you knew nothing of Mr. Hook's invention, before you sent your anagram."

[2] During all this time it must be remembered that active war was in progress with Holland. On August 8, 1667, Evelyn writes: "Visited Mr. Oldenburg, now close prisoner in the Tower, being suspected of writing intelligence. . . . This gentleman was Secretary to our Society . . . and I am confident will prove an innocent person." He was detained for two months. He was subsequently reinstated as Secretary until his death in 1678. He had visited Leiden in 1661, when no doubt he discussed horological questions with Huygens.

to Huygens. The implication appears to be that Oldenburg, from his official position in the Society and his intimate relations with Lord Brouncker, the President, Moray, Boyle, etc., had acquired knowledge of what Hooke had "revealed" in 1660 and communicated it to Huygens. There appears to be a veiled insinuation against these "persons of honour."

Birch ("History of the Royal Society," vol. iii, p. 324) chronicles a "debate concerning a scandalous postscript annexed to a book called Lampas," and a committee was appointed to bring in a Report.

This called forth, in the Transactions of November 30, 1676—

A Declaration of the Council of the Royal Society, passed Novemb: 20, 1676, relating to some passages in a late Book of Mr. Hooke entitulated "Lampas &c." reflecting on the integrity and faithfulness of the publisher in his management of the intelligence of the Society: the Council hath thought fit to declare in behalf of the publisher that they knew nothing of the publication of the said book: and further that the said publisher hath carried himself faithfully and honestly in the management of the intelligence of the Royal Society and given no just cause for any such reflections.

Whether Hooke had really invented something superior to Huygens' mechanism—he says in his Postscript to "Helioscopes" that "he was unwilling to add any of the better applications of the springs, waiting for an opportunity more to his advantage"—can never be known; for his secret, if any, died with him. Or can it be that the Tompion watch of 1675 embodied this improvement on Huygens' design, which, as Derham tells us, he did not think it worth the while to patent?

Thomas Reid (*op. cit.*, p. 265), in his defence of Hooke's claim, refers to a large collection of very valuable MS. letters between Huygens, Sir Robert Moray, Lord Brouncker and others, of which he was given the perusal in 1782 by a member of the Kincardine family, some of which are concerned with " what they called pendule watches," and deal with patents to be taken out in various countries and their respective share of profits. Reid concludes that these negotiations were the outcome of the Agreement as to the balance spring which Hooke declined to execute, and he therefore places these " persons of honour " in a very dishonourable light.

The original letters, which were in the possession of the late Lord Elgin, have so far not come to light ; but a transcript of them exists in the possession of Mr. William Douglas of Edinburgh. Having been favoured with extracts from some of them, I am able to state that they do not refer to the balance spring ; but to Huygens' " pendule " clock as modified for use at sea by Lord Kincardine.

I have explained that the going train of a clock was then called the " watch " ; whilst a pocket watch with a straight spring was known as a " pendule watch " ; hence Reid's grievous mistake as to the subject-matter of the Kincardine letters. I am glad that a prolonged search for this correspondence has enabled me to dispose of an unwitting calumny.

The subsequent history of the balance spring has been fully described in Lieut.-Commander R. T. Gould's " The Marine Chronometer," pp. 158–71.

THE CLOCKS OF JAPAN

LADY AND CLOCKMAKER.

From a Japanese Colour Print

INTRODUCTION

THE subject of Japanese horology is one which has received scant attention, although it is one which well deserves study, from the point of view both of the problems which the division of time in Old Japan presented for solution, and of the variety of design and artistic execution of Japanese clocks, especially during the latest period of the art.

After the revolution of 1866, which saw the end of the civil rule of the Tokugawa dynasty of Shōguns and the restoration of the Mikado, Japan threw itself open to Western influences and civilization ; and it was only to be expected that, amongst other drastic changes, the ancient and complex method of reckoning time would be abandoned. And so it came about that on January 1, 1873, European time-keeping and our calendar were finally introduced, and therewith all the existing clocks and timepieces became obsolete and useless, and found their way into the hands of curio dealers, and thence into the possession of European and American buyers.

The result has been that Japan has become largely denuded of these interesting objects ; and even the cultivated Japanese of to-day appear to be ignorant of the existence of these relics of pre-revolution days.

This is, however, the less surprising, because the use of clocks would appear to have been limited to the temples, and to the nobles and the wealthy, who could afford to employ a clockmaker to keep them adjusted, as the Japanese system of time measurement required.

Another reason for their scarcity may be attributed to the ordinary construction of Japanese houses and to their furnishing. A house built without solid walls, and with rooms divided by movable partitions, would necessarily offer no opportunity for setting up heavy weight clocks or brackets. The absence of heavy pieces of furniture, such as cabinets, chests of drawers, mantelpieces and so forth, would also preclude the use of spring-driven or " bracket " clocks in the way to which we are accustomed. Tall clocks, in the style of our " long-case " clocks, with the dial six feet from the floor, would also be unsuitable in a country where it was the custom to sit on the ground.

In a little treatise on the ceremonial of tea-drinking in Japan [1] an interesting comparison is drawn between the widely divergent Japanese and European conceptions as to the display of objects of art and furniture :

> The tea-room [says the writer] is absolutely empty, except for what may be placed there temporarily to satisfy some æsthetic mood. Some special art object is brought in for the occasion, and everything else is selected and arranged to enhance the beauty of the principal theme. One cannot listen to different pieces of music at the same time, a real comprehension of the beautiful being possible only through concentration upon some central motive. Thus it will be seen that the system of decoration in our tea-rooms is opposed to that which obtains in the West, when the interior of a house is often converted into a museum.
>
> To a Japanese, accustomed to simplicity of ornamentation and frequent change of decorative method, a Western interior permanently fitted with a vast array of pictures, statuary and bric-à-brac gives the impression of mere vulgar display of riches. It calls for a mighty wealth of appreciation to enjoy the constant sight of even a masterpiece, and limitless indeed must be the capacity for artistic feeling in those who can exist day after day in the midst of such confusion of colour and form as is often to be seen in the homes of Europe and America.

A clock, in the nature of things, called for a permanent position amongst the household gods, and it is easy to understand that a Japanese, holding such views as these, would not readily admit its presence. To the same reason may, perhaps, be attributed—apart from its simplicity and cheapness—the comparatively common occurrence of the small, vertical, so-called " pillar " clock, which could be

[1] " The Book of Tea," by Okakura Kakuzo, Edinburgh, 1919, p. 80.

attached to the principal pillar of the house without causing any appreciable interference with the austerity of its surroundings.

Finally, in a land where the " hours " varied with the season of the year and railways were unknown, the absence of a watch or timepiece would not make itself constantly felt.

In this connexion I would also quote a remark, which occurs in the late Sir Ernest Satow's book (p. 229).[1] He says :

> In those days [i.e. the years immediately preceding the Revolution] neither clocks nor punctuality were common ; and, as the hour altered in length every fortnight, it was very difficult to be certain about the time of day, except at sunrise, noon, sunset and midnight.

Sir Ernest Satow, who was attached to the British Mission during the momentous years 1866–1869, enjoyed exceptional opportunities for studying the social life and customs of the Japanese. Speaking and writing the language fluently, he was on terms of intimacy with many of the leading men of the day, and was permitted to travel in the interior and visit the castles of the great nobles at a time when the country was generally closed to European visitors. His testimony is therefore of unusual value. In reply to an inquiry of mine he was good enough to inform me that he never remembered seeing a clock in any Japanese interior. His Japanese servant, he added, could only recollect a clock in the temple of his village, and they depended on the sun for estimating the approximate time of the day. It must, however, be borne in mind that it is contrary to the custom of the Japanese to admit visitors to their private apartments. Guests would therefore only have access to the special rooms reserved for their entertainment, from which it is conceivable that the Japanese, with

[1] " A Diplomat in Japan : an Inner History from Personal Experiences of the Overthrow of the Shōgun Rule," London, 1921.

their innate sense of politeness, would rigidly exclude an instrument for marking the flight of time.

I have had the opportunity of exhibiting specimens from my collection on various occasions to members of the Embassy in London, and have always been met with the remark that they had never seen and were ignorant of the existence of any such clocks in their own country. In these circumstances it is not to be wondered at that little information as to Japanese horology is to be gleaned from the Japanese themselves, and that we are thrown back on the clocks that have survived for any evidence as to their evolution and age.

A book by Mr. Hyōé Takabayashi, published in 1924, entitled "Tokei Hattatsu-shi," gives a history of time measurement, and is illustrated with 76 plates. I refer to this treatise in the "Conclusion" to this section.

In 1796 Hanzo Hosokawa published a work in three volumes dealing with mechanical toys and clocks, one volume being entirely devoted to the latter. It describes all the parts and the construction of a single-balance "lantern" clock, fully illustrated in diagram form.

When speaking of clocks, it must be understood that our inquiry is limited to mechanisms whose motive power is derived from a falling weight or a compressed spring. The knowledge and use of such time-measurers was entirely derived from Europe. It is essential to remark that they were equally unknown to China, whence Japan derived, in the days of its early history, all its ideas as to the arts and sciences, including astronomy, and indeed all its literary cultivation.[1] The Chinese form of writing was adopted, and remains to this day in Japan as the written and printed character in general use. It is true that, side by side with the Chinese picture-characters, or ideograms, there is a

[1] Sundials were employed from very early times in China, and the use of water-clocks is believed to have been introduced from Greece. The burning of tapers or candles for measuring time was also practised.

native Japanese writing, based on pronunciation ; but with this we need not concern ourselves, the Chinese signs of the zodiac and the Chinese numerals being invariably used to indicate the " hours " on Japanese clocks.

The Portuguese discovered Japan about 1542, and were allowed to trade in the ports of Kiū-Shiū, and Francisco de Xavier arrived as a missionary in 1549. Objects of Western manufacture naturally created great curiosity, amongst them being fire-arms, a novel form of weapon which excited much interest amongst a warlike people. It is quite probable that clocks were included amongst the future articles of trade ; but, as the Portuguese were not clockmakers, we may suppose that they would have been of Italian, Dutch or German origin. The proselytizing activities of the Portuguese and Spaniards led eventually to wholesale massacres and to their final expulsion from Japan in 1638. But meantime Dutch traders had arrived on the scene. Iyeyasu, the founder of the Shōgun dynasty of the house of Tokugawa in 1598, was entirely favourable to foreign commerce, and the ports of Japan were freely open to European vessels ; if Christianizing propaganda had been left alone, it is probable that they would never have been closed.

In 1605 Iyeyasu granted a trading licence to the Dutch. This was followed by a charter to English traders from the East India Company in 1613 ; but after ten years' competition with the Dutch they abandoned the enterprise. Had the Dutch been content to restrict themselves to trading activities, they might have maintained their privileged position ; but they utilized the port of Hirado, which they had selected, for engaging in hostilities against the Spanish and Portuguese settlements in the East.

Iyeyasu died in 1616 and his successor did not entertain the same friendly disposition towards foreign trade and intercourse. As a result of various disputes the Dutch were finally relegated to the occupation of the tiny island of

Déshima, close to Nagasaki, where they were allowed to remain until Japan re-opened relations with the West in the nineteenth century.

Sir Ernest Satow, in his book previously referred to, mentions (p. 204) that the head of the Dutch trading factory was allowed periodically to travel from Nagasaki overland to Yedo (now Tokyo) to pay his respects to the Shōgun, and to convey valuable presents to him and his ministers.

The same author, in an article on "The Church of Yamaguchi from 1550 to 1586,"[1] quotes Xavier as saying that on returning to Yamaguchi (i.e. from Kiōto in 1550) he presented a clock from the Portuguese Viceroy of India, which, though of little value, was highly appreciated by the King (*sic.*), as something that had never before been seen in his province.

Mr. Takabayashi (*op. cit.*) confirms this statement by quotation from a biography of Yoshitaka Ouchi, the Governor of Yamaguchi in Suwo. The biography is attributed to the sixteenth century, and relates that Xavier, on applying to Ouchi for permission to start a mission, presented him with a clock. It is added that "among several gifts from India an instrument was included which controls the twelve 'hours' and rings during day and night."

Takabayashi's next reference to a clock is that of one given to Iyeyasu by a missionary, with a Latin inscription stating that it was made in Madrid in 1581. Fig. 24 in his book shows a four-square spring clock, with open-work dome to contain the bell, of that period.

Finally, he tells us that in 1591 four Japanese envoys returned from a mission to Europe. They were accompanied by a missionary and proceeded to Kyōto, where the latter presented Hideyoshi with a clock.[2]

[1] Transactions of the Asiatic Society of Japan, vol. vii, p. 135. M. Planchon ("L'Horloge," Paris, 1898, p. 210) erroneously gives the reference as 'Satorr. (t. vii, p. 2)."

[2] Hideyoshi, the great general and statesman, having finally crushed the centuries-long feuds among the daimyōs, rose to supreme power and became civil ruler (Shōgun) of all Japan. At first favourable to Christianity, he

We may suppose that the Dutch traders would also include European clocks amongst their offerings, and that in this way the Japanese, during all the long period of the closing of their country against Europeans, would be made acquainted with all the phases of horological development in the West.

The late M. Planchon (*op. cit.*) refers to the " first attempts at manufacture made by the natives under the inspiration and direction of Europeans." It is quite possible that amongst the Jesuit fathers there were some who were skilled in clockmaking ; but, however this may be, it is certain that the crude domestic clock of the sixteenth century was slavishly copied in Japan. The great skill of the Japanese in the working of metals made them peculiarly adapted for practising this new art.

A problem at once presented itself, however, for solution. The European clock was based upon a system of twice twelve hours of equal duration, struck in ascending progression from 1 to 12 ; whereas the Japanese, adopting the ancient Chinese system, reckoned time according to the " natural " day, the length of its division varying with the seasons, marked by strokes on the temple bell in descending order from 9 to 4. If then we find a Japanese clock of archaic construction—of which my collection contains a specimen—which strikes European hours, in conformity with the notches in its original locking plate, it is fair to assume that it was either the work of a European clockmaker in Japan, or a slavish Japanese copy, made before the possibility of adapting the mechanism to the native system of time measurement had been appreciated.

decreed stringent measures against the Jesuits in 1587 ; but in the following year found it politic to treat their presence with toleration, until a year before his death in 1598 he reverted to suppression and executions. It was therefore during this interval of leniency that he accepted the clock.

CHAPTER I

JAPANESE TIME MEASUREMENT

BEFORE proceeding to any description of Japanese clocks it is necessary to understand the way in which the Japanese divided the day. It had its origin in China ; and, as stated in the Introduction, was based upon the natural day, but commencing at dusk. Each period from dusk to dawn and from dawn to dusk was divided into six *equal* divisions ; but, inasmuch as the length of each period varied with the seasons, it follows that the duration of the divisions was greater or lesser according to the time of the year.

It will thus be understood that, if for convenience we speak of " hours " in connexion with Japanese clocks, their duration is not only not one of sixty minutes, but also one of constantly varying length.

Each of these divisions, beside having its numeral attached to it, corresponding with the number of strokes on the bell, was also associated with one of the Chinese signs of the zodiac, according to the following scheme :

	Numeral.	*Sign of Zodiac.*		
6.	六	酉	Cock	Dusk
5.	五	戌	Dog	
4.	四	亥	Boar	
9.	九	子	Rat	Midnight
8.	八	丑	Bull	
7.	七	寅	Tiger	

	Numeral.	*Sign of Zodiac.*		
6.	六	卯	Hare	Dawn
5.	五	辰	Dragon	
4.	四	巳	Serpent	
9.	九	午	Horse	Midday
8.	八	未	Goat	
7.	七	申	Ape	

On the earlier clocks, with a fixed "hour-circle" and revolving hand, the Chinese zodiacal signs are inscribed in an outer ring, with the corresponding numerals below them in an inner ring. The Horse sign for noon appears at the top, with the Rat for midnight opposite to it at the bottom; and, as the hand revolves from left to right, the Cock for dusk occupies the central position on the right and the Hare for dawn on the left of the hour circle.

With the introduction of the revolving hour-circle with its small adjustable "hour" plates, space only admitted of the marking of the numerals; but on the "pillar" clocks the plates were occasionally inscribed with zodiacal signs only; whilst on some of them the signs for noon and midnight were retained, the remainder of the plates bearing the numerals.

The reason for employing the astronomical signs as well as the numerals was doubtless because each "hour" was designated by its zodiacal name. It has also been suggested to me that, as certain hours were lucky whilst others were unlucky, a glance at the clock would at once show whether the time for an action was propitious or not.

According to the ancient Chinese system of reckoning,

the zodiacal sign represented the middle of the "hour" period to which it was assigned. Thus, supposing the division to be equal to two of our hours, the Horse "hour" would have begun at 11 a.m. and have ended at 1 p.m., when the Goat "hour" would commence.

The Japanese abandoned this system and reckoned the beginning of the "hour" from the tolling of the temple bell or the striking of the clock, so that in this case the Horse "hour" lasted from noon to our 2 p.m.

It will be observed that the Japanese, whose methods are in so many instances the reverse of those of the West, counted the hours backwards. The highest number, 9, was allotted to midday and midnight ; and as they only required six numerals for each of the two periods of the day, the progression of the "hour" is represented by 9, 8, 7, 6, 5, 4.

Several explanations of this system have been advanced ; but I am not able to say that any one of them is correct. It seems certain, however, that the numbers 1, 2 and 3 were reserved for temple use. When clocks came into use, they must necessarily have conformed to the time-honoured custom of the temple, which signalled the successive hours to the community by strokes on the great bell.

According to Hoffmann [1] the practice at Yedo (Tokyo), and doubtless generally, was as follows : first a single stroke was given to the bell, then at an interval of about a minute a second blow, followed quickly by a third. There was then another long pause, and finally the number of strokes corresponding with the hour was sounded, with an interval of about ten seconds between each stroke, except the last two, the final stroke following rapidly, in order to indicate that the full number was completed.

To return to the numbers : one theory is that the numerals

[1] Dr. J. J. Hoffmann, "Japanische Sprachlehre," 1877, German translation from the Dutch. He devotes several pages to time measurement and the calendar. See also P. F. von Siebold's great work on Japan : "Nippon ; Archiv zur Beschreibung von Japan," 5 vols.

from 1 to 6 were multiplied by the mystic number 9, and that the second figure of the product was adopted, thus:

1	2	3	4	5	6
9	9	9	9	9	9
9	18	27	36	45	54

On the other hand, as the "hours" were divided decimally the numbers 1 to 6 may have been deducted from 10:

10	10	10	10	10	10
1	2	3	4	5	6
9	8	7	6	5	4

However this may be, it seems probable that the number 9, with its supposed magic properties, was selected to indicate the one invariable moment of the day when the sun was in the meridian, instead of applying it, as the first number of the descending series, to the ever-changing commencement of the natural day at dusk.

Another, somewhat fanciful, reason has been suggested. The working population of Japan was in old days for the most part engaged in agriculture and fishing. Rising at dawn, the period of labour still left to be accomplished until the noonday rest grew less with each succeeding hour, which the diminishing numbers expressed. After midday the same downward progression continued until 6 was reached, when darkness set in.

The suggestion may sound ingenious; but one may be permitted to ask how the fisherman at sea, or the labourer on a distant farm, far from his village, was supposed to hear the tolling of the temple bell. Nor is it likely that the humble toiler was furnished with a pocket sundial, such as was used in China.[1]

[1] These sundials are made of two flat pieces of wood, hinged together. The lower piece contains a compass, and when the upper piece is raised to a right angle, a cord stretched between the two, acting as a gnomon, throws a shadow on the dial on the lower piece.

The name given to the "hour" period is *toki*, and each *toki* was decimally divided into 10 *buns*. For accurate measurement the *bun* was divided into 10 *rin*. On clock dials the *toki* has a half-"hour" division with five tenths on either side, but space does not admit of any further subdivision.

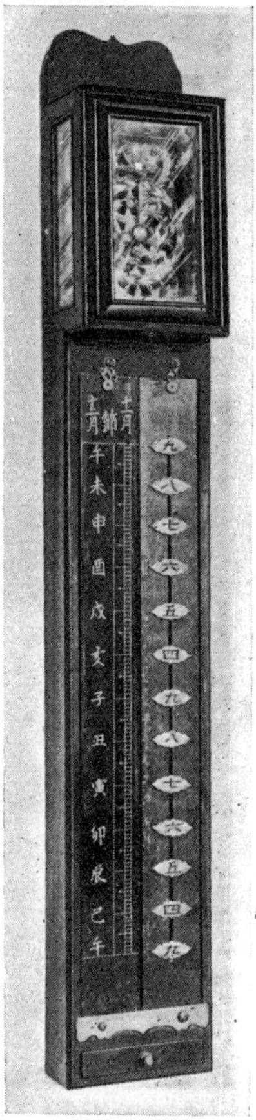

Fig. 1.

The division of the "hour" into tenths served a useful purpose so long as the hand travelled round the inscribed "hour" circle; but it ceased to indicate a division of the period when clocks with shifting "hour" plates and a stationary hand were introduced, or when, as in the case of the long "pillar" clock, the hand was attached to the descending weight. In the former case the tenths were engraved on the outer rim of the "hour" circle; in the latter case they were commonly cut on a brass scale, screwed to the side of the wooden case. They were therefore immovable. One must therefore suppose that they were subdivisions of measure and not of time, and that their purpose was to guide the clockmaker when at the end of each half-month he came to adjust the clock to the time for the succeeding half-month, by shifting the movable plates nearer together or farther apart so as to correspond with the altered length of the day and night hours.

Whilst division into tenths was the general practice, I may mention that one unusually long pillar clock in my collection (Fig. 40) has each tenth subdivided into thirtieths.

Before leaving the subject of time measurement I would

remark that, in the case of some pillar clocks, with the lacquer "hour" scales, the uppermost figure or zodiacal sign is not that which corresponds with nightfall, as is usual, but the Horse or midday. It may be that such scales date from the introduction of European time, so that the clock could still be used for the new system by spacing all the numerals equally (Fig. 1).

Clocks of the pillar type are of frequent occurrence, and, in their simplest form, must have been quite cheap, and therefore within the means of the poorer classes. Moreover, when they were furnished with a set of scales giving the "hours" and tenths according to the progression of the seasons, the fortnightly services of a skilled clockmaker to adjust the "hour" plates were unnecessary. But under the new system it is easy to understand that the fixed hour of noon would be a more convenient time than nightfall for winding, bearing in mind that, with rare exceptions, no Japanese clock was made to go for more than a day.

These simple clocks, as I have mentioned, took up no room, and were hung to the central pillar of the house.

Beside indicating the time, most of the weight clocks, except the earliest, and almost all of the spring-driven clocks also gave the date, as explained in the next chapter. This improvement was doubtless suggested by clocks introduced from Europe, where the practice became general towards the close of the seventeenth century. The calendar system is explained in the next chapter.

CHAPTER II

JAPANESE CALENDARS

LIKE European clocks, the more elaborate Japanese clocks not only give the time of day and strike the hours and half-hours, and sometimes possess an alarum, but they also indicate the date. This is shown by signs which appear in small rectangular openings usually below and on either side of the hour-circle. These signs are engraved on two wheels fixed behind the front plate, one having twelve teeth and the other ten, which engage in one another, and from the motion work are made to move forward one tooth every twenty-four hours. By this means any two given signs only come together once in sixty days. The signs on the larger wheel on the left are the twelve signs of the Chinese zodiac, whilst those on the smaller wheel represent the five elements and sub-elements.

It is unnecessary here to explain at any length the very complicated methods by which the Japanese, following the older Chinese systems, determined chronological dates ; but it may be mentioned that these methods were three in number.[1] The oldest is that of counting by the years of the reigns of the Emperors. A parallel to this still exists in this country, where Acts of Parliament are dated by the year of the sovereign's reign. The second method is that of reckoning by periods or epochs of uncertain length, dating from some important event after which the period (called *nen-gō*) is named. This is as if we commenced a period with the Battle of Waterloo or from August 4, 1914 ; but the beginning of the period was dated *back* to the commencement of the year when the event happened. In this way the first year of the Waterloo period would be

[1] An exhaustive treatise on Japanese chronology and calendars will be found in the *Transactions of the Asiatic Society of Japan*, Supplement of vol. xxxvii, 1910, by Prof. E. W. Clement and William Bramsen, from which much of the following information is derived.

reckoned from January 1, 1815. If the next event chosen to represent a period were the passing of the Reform Bill, the Waterloo period would have seventeen years, and the Reform period would commence its first year on January 1, 1832, and so on.

On the other hand, the date of the Emperor's reign was usually reckoned from the commencement of the year *following* his accession. Following up our parallel, we should consider the reign of George III as beginning on New Year's Day, 1761, and lasting until New Year's Day, 1821, so that the year 1820 would be called the sixtieth year of George III and the sixth year of the Waterloo period.

The third system originated in China in 2637 B.C., and is termed the "sexagenary cycle," that is, a cycle of sixty years. The years of a cycle are designated by combining the names of the signs of the zodiac (called the Twelve "Terrestrial Branches") with the signs for the particular element (called the Ten "Celestial Stems"). This sexagenary cycle was similarly used to designate cycles of months and days, and hence we see its application to the date indicated by the clock.[1]

The names of the signs of the zodiac denominating the months and hours are in Japanese as follows :

	Month.			*Hour.*	
1.	子	ne	Rat (winter solstice)	Corresponding with	9 Midnight
2.	丑	ushi	Bull	do.	8
3.	寅	tora	Tiger	do.	7

[1] T. S. Bayer, in a brochure published in St. Petersburg in 1735, a copy of which exists in the Library of the Patent Office, describes the method of time measurement employed by the Chinese, and shows that the signs of the zodiac, corresponding with the periods of time, were those adopted by Japan, although the written characters are different. He states that the same animals were represented in the Persian system, except that the crocodile took the place of the dragon, and he argues that both systems were themselves derived from a common astronomical source.

	Month.				Hour.
4.	卯	u	Hare (vernal equinox)	Corresponding with	6 Dawn
5.	辰	tatsu	Dragon	do.	5
6.	巳	mi	Serpent	do.	4
7.	午	uma	Horse (summer solstice)	do.	9 Noon
8.	未	hitsuji	Goat	do.	8
9.	申	saru	Ape	do.	7
10.	酉	tori	Cock (autumnal equinox)	do.	6 Dusk
11.	戌	inu	Dog	do.	5
12.	亥	i	Boar	do.	4

The Elements, or Celestial Stems, are the following :

1.	癸	ki-no-e	Wood	(Elder brother)
2.	壬	ki-no-to	,,	(Younger brother)
3.	辛	hi-no-e	Fire	(Elder brother)
4.	庚	hi-no-to	,,	(Younger brother)
5.	己	tsuchi-no-e	Earth	(Elder brother)
6.	戊	tsuchi-no-to	,,	(Younger brother)
7.	丁	ka-no-e	Metal	(Elder brother)
8.	丙	ka-no-to	,,	(Younger brother)
9.	乙	mizu-no-e	Water	(Elder brother)
10.	甲	mizu-no-to	,,	(Younger brother)

It will be observed that the element in its primary state always ends in "e," whilst in its secondary state, as fashioned or utilized by man for his own use, it has a termination in "to." These two endings, forming the word E-to, are used to designate a combination of a celestial stem with a terrestrial branch corresponding with the particular ordinal number of the sexagenary cycle of years or months or days. In giving the Eto the name of the Element precedes that of the zodiacal sign. Thus the first Eto of a cycle is called ki-no-e ne ; the second ki-no-to ushi ; the eleventh is ki-no-e inu ; the thirteenth hi-no-e ne. The complete progression of the cycle as it appears on the clock is shown on p. 208, the "elder brother" being in italics.

The earliest Japanese calendars, based upon their Chinese prototype, divided the year into lunar months. The length of a lunation is taken as being 29 d. 12 h. 44 m. 3 s., according to our mode of reckoning ; but, as the length of a day is obviously determined by the sun, the length of the lunar month was reckoned as containing either 29 or 30 solar days. If a lunation were exactly 29½ days the problem would have been easily solved by employing alternate months of 29 and 30 days. But the additional 44 minutes 3 seconds necessitated a periodical adjustment. The method by which this was attained was as follows [1]:

The day for this purpose was deemed to commence at midnight, and the lunation was reckoned from new moon to new moon. When the sum of two lunations, after deducting the number of days allotted to the preceding month, was less than 30, the month was considered to contain 29 days ; but if the result exceeded 30, then the month was reckoned as one of 30 days.

[1] For the sake of simplicity I make use of the European system of equal hours of sixty minutes, omitting the seconds. The length of a lunation according to certain Chinese calendars is expressed decimally by Mr. Bramsen as 29·5305921 days, which agrees with the actual length of a lunation, namely, 29 d. 12 h. 44 m. 3 s.

1.	甲 子	*wood*-rat	21.	甲 申	*wood*-ape	41.	甲 辰	*wood*-drag
2.	乙 丑	wood-bull	22.	乙 酉	wood-cock	42.	乙 巳	wood-serp
3.	丙 寅	*fire*-tiger	23.	丙 戌	*fire*-dog	43.	丙 午	*fire*-horse
4.	丁 卯	fire-hare	24.	丁 亥	fire-boar	44.	丁 未	fire-goat
5.	戊 辰	*earth*-dragon	25.	戊 子	*earth*-rat	45.	戊 申	*earth*-ape
6.	己 巳	earth-serpent	26.	己 丑	earth-bull	46.	己 酉	earth-cock
7.	庚 午	*metal*-horse	27.	庚 寅	*metal*-tiger	47.	庚 戌	*metal*-dog
8.	辛 未	metal-goat	28.	辛 卯	metal-hare	48.	辛 亥	metal-boa
9.	壬 申	*water*-ape	29.	壬 辰	*water*-dragon	49.	壬 子	*water*-rat
10.	癸 酉	water-cock	30.	癸 巳	water-serpent	50.	癸 丑	water-bull
11.	甲 戌	*wood*-dog	31.	甲 午	*wood*-horse	51.	甲 寅	*wood*-tiger
12.	乙 亥	wood-boar	32.	乙 未	wood-goat	52.	乙 卯	wood-hare
13.	丙 子	*fire*-rat	33.	丙 申	*fire*-ape	53.	丙 辰	*fire*-dragon
14.	丁 丑	fire-bull	34.	丁 酉	fire-cock	54.	丁 巳	fire-serpen
15.	戊 寅	*earth*-tiger	35.	戊 戌	*earth*-dog	55.	戊 午	*earth*-horse
16.	己 卯	earth-hare	36.	己 亥	earth-boar	56.	己 未	earth-goat
17.	庚 辰	*metal*-dragon	37.	庚 子	*metal*-rat	57.	庚 申	*metal*-ape
18.	辛 巳	metal-serpent	38.	辛 丑	metal-bull	58.	辛 酉	metal-cock
19.	壬 午	*water*-horse	39.	壬 寅	*water*-tiger	59.	壬 戌	*water*-dog
20.	癸 未	water-goat	40.	癸 卯	water-hare	60.	癸 亥	water-boar

Note.—Some clocks have only a single date indicator, which in this case is the cycle of the signs of the zodiac.

If we commence with a new moon at the beginning of a sexagenary cycle at 0, we obtain the following result :

First month	= 29.12.44		29 days
Second month, difference	12.44	= 30.1.28	30 ,,
Third month, difference	1.28 29.12.44	= 29.14.12	29 ,,
Fourth month, difference	14.12 29.12.44	= 30. 2.56	30 ,,
Fifth month, difference	2.56 29.12.44	= 29.15.40	29 ,,
Sixth month, difference	15.40 29.12.44	= 30. 4.24	30 ,,

and so on.

By continuing this process, we shall find that the sixteenth month works out at 30 d. 11 h. 44 m., and the seventeenth month at 30 d. 0 h. 24 m., so that to each was allotted 30 days. A long month following a long month recurs at each interval of seventeen months, that is to say, that in the cycle the long months were duplicated at the seventeenth, thirty-fourth and fifty-first months, whereas two short months should never fall together. But this rule was not adhered to in later times.

Side by side with the calendar based upon lunar months, the Japanese also employed a system founded on a solar year of 365 days. It is obvious that the discrepancy between the lunar and solar months increased rapidly, and adjustment by means of an intercalary month in the lunar calendar became periodically necessary.

If we take 33 months of the lunar calendar, as above described, we shall find that it contains 16 months of 29 days and 17 months of 30 days, giving a total of 974 days. On the other hand, by dividing 365 by 12 and multiplying by 33 we arrive at 1,003¾ days, or a difference of 29¾ days. In order to redress this error, and to bring the lunar calendar periodically into approximate harmony with the solar year, it was the practice to insert an intercalary month, sometimes of 29 days, and sometimes of 30 days, at the end of a period approximating 33 lunar months.

Before proceeding, however, to show how the civil year, based on lunar months, was made to correspond from time to time with the solar year, it is necessary to explain the latter system.

The year was divided into twelve parts or months, termed *setsu*, each of them having its special designation. The length of a setsu varied from 29 to 32 days, according to the period between the equinoxes, and each setsu was again approximately divided into two, the half-period being known as chū-setsu. (中)

The period between the commencement of one chū-setsu and that of the succeeding one corresponded theoretically with the divisions of the year according to the zodiacal signs or Terrestrial Branches. Thus the first chū-setsu began at the date of the winter solstice, which was reckoned to correspond with December 22 of our calendar, the first month of thirty days being that of the Rat, and so on, following the order of the zodiacal signs already given. Nevertheless, as the chū-setsu is a division of the setsu, the setsu immediately preceding the winter solstice was regarded as the actual commencement of the *solar year* (December 8).

In the later calendars account was taken of the fact that the period between the vernal and autumnal equinoxes exceeds that between the autumnal and vernal equinoxes, the dates corresponding with March 21 and September 23 in our calendar, the months from spring to autumn containing 186 days, whilst those from autumn to spring number only 179.

The civil year, as stated, was composed of lunar months ; but here a further difficulty is encountered, inasmuch as it did not commence at the same time as the solar year. In theory it was supposed to begin with the third setsu

(February 4) ; but, owing to the ever-varying discrepancy arising from the difference in lunar and solar months and years, the civil year in practice might commence as early as *our* 20th of January or fall as late as *our* 22nd of February.

The discrepancy between the lunar and solar systems of reckoning was adjusted, as I have stated, by the insertion at intervals of an intercalary month. The time at which this took place was not arbitrary, but was determined by a rule which may be stated in this way.

Let us suppose that we start with a cycle in which the new moon did in fact correspond with midnight at the commencement of the third setsu of the solar year on February 4. It is obvious that the next new moon would fall in advance of the following or fourth setsu or *solar* month. As this error is augmented during the succeeding months, the difference will at last become so great that the new moon will outstrip the chū-setsu or half-solar month, and the completion of the next lunation will be reached before another chū-setsu has arrived. Such a lunation, taking place without the occurrence of a chū-setsu during its course, was adopted as the intercalary month. This month bore the name of the month which it followed with the word Urō prefixed.

If we take the length of a *solar* month (one-twelfth of 365 days) as being 30 d. 10 h. and that of the lunar month as 29 d. 12 h. 44 m. 3 s., we find that the difference is one of 21 h. 15 m. 57 s. Multiplying this difference by 33 the result is found to be 29 d. 5 h. 46 m. 21 s., thus proving in another way that the interposition of an intercalary month, sometimes of 29 and sometimes of 30 days, at intervals of about 33 months will approximately restore harmony between the civil and the solar years.

The Japanese would appear to have been acquainted with the Metonic cycle of nineteen lunar years, a period at the end of which the moon returns to the same apparent position in relation to the sun, so that the new moon occurs at the same date in the corresponding year of each cycle. This cycle contains 235 lunations. For the purpose of illustrating the adjustment of the civil and solar years by intercalary months, let us take the Metonic cycle commencing in 1800 and terminating in 1819. The intercalary months are placed in heavy type :

Sexagenary cycle		*Months*	*Interval*	*Days*
57 [1800]	.. .. 30, 29, 30, 29, 30, 29, 30, 29	**8**		236
58 [1801]	30, 29, 30, 29, 30, 29, 30, 30, 29, 30, 29, 30	12		355
59 [1802]	29, 30, 29, 29, 30, 29, 30, 30, 30, 29, 30, 29	12		354
60 [1803]	30, **29**, 30, 29, 29, 30, 29, 30, 30, 29, 30, 30, 29	**13**	33	384
1 [1804]	30, 29, 30, 29, 29, 30, 29, 30, 29, 30, 30, 30	12		355
2 [1805]	29, 30, 29, 30, 29, 29, 29, 30, **29**, 30, 30, 30, 29	**13**	31	383
3 [1806]	30, 30, 29, 30, 29, 29, 29, 30, 29, 30, 30, 29	12		354
4 [1807]	30, 30, 30, 29, 30, 29, 29, 30, 29, 29, 30, 30	12		355
5 [1808]	29, 30, 30, 29, 30, 29, **30**, 29, 30, 29, 29, 30, 29	**13**	34	383
6 [1809]	30, 30, 29, 30, 30, 29, 30, 29, 30, 29, 30, 29	12		355
7 [1810]	29, 30, 29, 30, 30, 29, 30, 30, 29, 30, 29, 30	12		355
8 [1811]	29, 29, **30**, 29, 30, 29, 30, 30, 29, 30, 30, 29, 30	**13**	32	384
9 [1812]	29, 30, 29, 29, 30, 29, 30, 29, 30, 30, 30, 29	12		354
10 [1813]	30, 29, 30, 29, 29, 29, 30, 29, 30, 30, 30, **29**, 30	**13**	33	384
11 [1814]	30, 29, 30, 29, 29, 29, 30, 29, 30, 30, 29, 30	12		354
12 [1815]	30, 30, 29, 30, 29, 29, 29, 30, 29, 30, 29, 30	12		354
13 [1816]	30, 30, 29, 30, 29, 30, 29, 30, **29**, 29, 30, 29, 30	**13**	33	384
14 [1817]	30, 29, 30, 30, 29, 30, 29, 30, 29, 29, 30, 29	12		354
15 [1818]	30, 30, 29, 30, 29, 30, 30, 29, 30, 29, 29, 30	12		355
16 [1819]	29, 30, 29, 30, **29**	**5**	32	147
		235		6,939

The length of the months is calculated from their commencing dates as given in Mr. Bramsen's Chronological Tables, above referred to.

An analysis of the foregoing figures shows that the period in question does in fact contain 235 lunar months. But as it contains 6,939 days, the average length of a month works out at 29 d. 12 h. 39 m. 49·8 s., as against the true length of a lunation of 29 d. 12 h. 44 m. 3 s., that is to say, there is a deficiency of 4 m. 13·2 s. per lunar month. This deficiency only amounts to 16½ hours in the whole Metonic cycle, an error which would be approximately compensated by the addition of an intercalary day over a period of twenty-eight solar years, the length of a "solar cycle." Whether this was done in practice I am unable to say.

There is also a discrepancy in the length of the solar year, inasmuch as 19 years containing 6,939 days give an average duration of 365 d. 5 h. 3 m. 9·5 s. as against the true length of 365 d. 5 h. 48 m. 51·6 s., equal to a shortage of 45 m. 42·1 s. This error would be redressed by intercalating an extra day after 31·5 years.[1]

It must be borne in mind, however, that not every period of nineteen years contained, as this does, exactly 205 intercalary *days*.

It will also be observed that the sequence of long and short months does not correspond with the original Chinese scheme, explained on p. 204.

In addition to the calendar systems above described the solar year was popularly divided into twenty-four seasons or *sekki*, designated by picturesque names descriptive of the expected weather or of agricultural activities.[2] These are given by Hoffmann[3] as follows:

1.	立春	Ris-sun	Beginning of spring	= Feb.	3
2.	雨水	U-sŭi	Rain-water	= ,,	19
3.	驚蟄	Kiyoo-tsits	Awakening of insects	= Mar.	5
4.	春分	Sun-bun	Mid-spring	= ,,	20
5.	清明	Sei-mei	Bright	= Apr.	5
6.	穀雨	Kokŭ-u	Seed rain	= ,,	20

[1] Actually, the lunar cycle contains 16½ hours *more* than 6,939 days.

[2] Compare the French calendar of the Revolution.

[3] Dr. J. J. Hoffmann's "Japanische Sprachlehre," 1877, pp. 164 *et seq.* German edition of the Dutch work.

7.	立夏	Rik-ka	Beginning of summer	= May 5
8.	小満	Seo-man	Little overflow	= ,, 20
9.	芒種	Boo-siu	Planting of rice	= June 5
10.	夏至	Ge-zi	Midsummer	= ,, 21
11.	小暑	Seo-sǐyo	Little heat	= July 6
12.	大暑	Dai-sǐyo	Great heat	= ,, 23
13.	立秋	Ris-siu	Beginning of autumn	= Aug. 7
14.	處暑	Siyo-sǐyo	Local heat	= ,, 23
15.	白露	Fakŭ-ro	White dew	= Sept. 8
16.	秋分	Siu-bun	Mid-autumn	= ,, 23
17.	寒露	Kan-ro	Cold dew	= Oct. 8
18.	霜降	Soo-koo	Hoar frost	= ,, 23
19.	立冬	Rit-too	Beginning of winter	= Nov. 7
20.	小雪	Seo-sets	Little snow	= ,, 22
21.	大雪	Dai-sets	Great snow	= Dec. 7
22.	冬至	Too-zi	Mid-winter	= ,, 22
23.	小寒	Seo-kan	Little frost	= Jan. 6
24.	大寒	Dai-kan	Great frost	= ,, 20

Fig. 26 shows a " bracket " clock which, in addition to a double calendar, gives the above progression of the seasons, the illustration displaying Nos. 12 and 13 in the above table. The early clocks of the seventeenth century were not provided with calendar work, an improvement which may be ascribed to the next century, following its adoption in European clocks.

STRIKING-CLOCK AND STAND, WITH CHINESE SIGNS OF THE ZODIAC ON DIAL PLATE

CHAPTER III

JAPANESE CLOCKS[1]

Clocks of native workmanship, made to comply with the Japanese system of time measurement, may be divided broadly into three main categories :

1. Clocks driven by weights supported on a stand, which stood on the floor, or were placed on a bracket, or suspended by a loop.

2. Portable spring-driven clocks enclosed in small glazed wooden cases, to rest on a bracket, or on some piece of furniture.

3. Clocks enclosed in a narrow wooden case, which contains the weight, intended to be hung against the pillar of the house or other permanent support.

For convenience of reference I shall term the clocks belonging to Class 1 " lantern " clocks, on account of their general resemblance to our own old brass clocks, although in point of fact their construction was based on the Continental and not on the English model.

Class 2 I shall term " bracket " clocks, as being the equivalent in miniature of our own bracket clocks.

Class 3 has no exact European prototype, and I shall designate clocks in this category as " pillar " clocks.

Other clocks were made, which do not fall under any of the above classes ; but these must be regarded as special departures from the general type, and call for individual description.

" Lantern " Clocks

We have seen from the short historical survey in the Introduction that the first clocks that found their way to Japan were introduced by the Portuguese missionaries and

[1] In some cases it has been necessary to remove either the hood or the glazed front slide in order to display the movement. The height, when given, is the measurement over all. For the Japanese names see p. 287.

the Dutch traders, and one may suppose from the intimate connexion of the Jesuit mission with Rome that the clocks were either of Italian manufacture, or later on of Dutch origin. The general design, however, of domestic clocks had been standardized throughout the Continent in the sixteenth century ; so that for our present purpose the exact provenance of the model is immaterial.

It is probable that the few clocks that were first imported as gifts to the Emperor and the great nobles would have been regarded in the light of European curiosities, much as a Japanese clock is now regarded as a " curio " in Western countries. The hour signs on the dial conveyed as little meaning to the Japanese as do the Chinese characters on a Japanese clock to most people to-day ; whilst the strokes sounded on the bell bore no relation to the time signals of the temple.

It must have taken some little time before it was realized that these strange assemblages of toothed wheels and pinions could in fact be adapted with slight alteration to meet the requirements of the local time-system.

I am therefore inclined to think that the first attempts at clockmaking were not made until some time after the close of the sixteenth century. If, amongst the Portuguese or Spanish missionaries or the Dutch traders, there was no one skilled in the horological craft, the Japanese artificer must have required long practice before he became proficient in wheel-cutting, and lengthy study before he grasped the principles of the calculation of a train of wheels and pinions, and particularly of the striking mechanism.

It is a matter of common knowledge that the Japanese workman was and is a past master in the art of imitation. Given a Dutch clock, his earliest efforts would be expended upon producing an exact copy of every piece and every detail of design, and in course of time he would succeed in putting together a clock in every respect an exact replica of the original.

I am supposing that our Japanese mechanic would only replace the European hour circle and its Roman numerals by one giving the twelve Chinese signs of the zodiac and their corresponding numerals, whilst still retaining the locking plate with its notches for striking from 1 to 12. Such I take

Fig. 2.

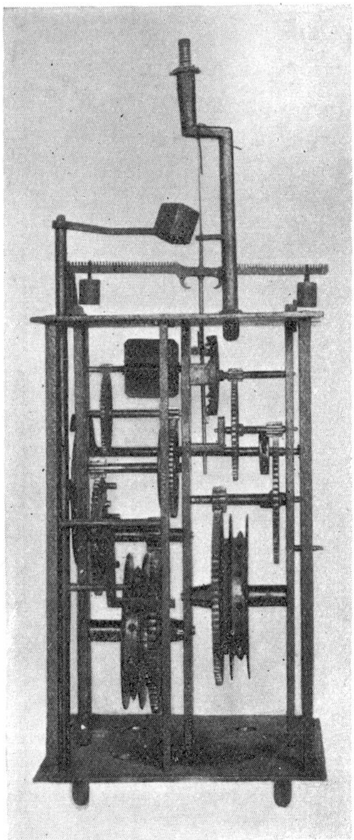

Fig. 3.

to be the origin of the clock which is illustrated in the frontispiece to this chapter (p. 216).

Figs. 2 and 3 represent the clock itself, and the archaic steel works, stripped of the outer casing. It measures 17 inches to the top of the bell.

The casing is ornamented with simple floral engraving, with the flower of the paulownia, the badge of the Shōguns of the House of Tokugawa, four times repeated on the front plate. Whether this is an indication that the clock was actually made for the reigning civil sovereign, and, if so, for which one, I am unable to say; but it is quite possible that it dates from the reign of Iyeyasu (1598–1616), the first of his line.

Fig. 4.

A clock almost identical in design is shown on this page, and it is also 17 inches in height (Fig. 4).

This clock is fitted with a locking plate which strikes the "hours" from nine to four in accordance with the Japanese system. Here then we get the first step forward in native clockwork.

The hour circle is of brass, and is of later date than the clock itself.

The cases of these early clocks were commonly made of copper plates silvered, of which Fig. 4 is one, and often without decoration. Specimens also occur in which the casing is of damascened iron or steel.[1]

This latter clock (Fig. 5), as the rarest exception, is inscribed beneath with the *date*, equivalent in our calendar to 1678. A similar clock in my collection is dated 1695.

[1] A very fine example of this work, with two balances, on its pyramidal stand, may be seen in the clock section of the Science Museum at South Kensington. Another specimen without stand is exhibited in the Victoria and Albert Museum.

The simplicity of the method of applying the casing to early clocks of this type is worthy of remark. No troublesome pins are used to hold on the front and back plates, as in our lantern clocks. The back plate is hinged at top and bottom with flat hinges, riveted to the side plates. When cutting the front and back plates two small tongues are left at the bottom in each, which drop into corresponding slots in the iron frame of the clock; the plates are then pressed back, and—so accurately are they cut—fit perfectly tight under the top of the frame. But for greater security they are fastened in front and at back by an ornamental screw. The side doors are then swung round, and a catch, with which each is provided, drops into a slotted projection on either side of the front plate. The revolving central disk carrying the hand or pointer is also held in place by a screw with a projecting head. All then that is necessary in order to remove the casing is to undo the three knobbed screws with the fingers, to raise the catches, to let the central disk fall out, whereupon the whole casing comes away—the work of less than half a minute (well shown in Fig. 23).

Fig. 5.

The central disk is the "dyal" or motion-wheel of Froissart's poem, with this difference. In de Vick's clock the hours were painted on the wheel itself, each hour number, as the wheel revolved, coming in front of a fixed index. This primitive method was discarded in Europe when the fixed hour circle with its numerals was introduced. The motion-wheel then carried with it a pointer, or hand, made fast to the wheel, which therefore travelled round the fixed hour circle. But, in order not to expose the teeth of the motion-wheel, a disk, fitting just inside the hour circle, was attached to the wheel to act as a cover. Such was the type of dial employed in the old European clock which the Japanese first copied, as may be seen in the illustrations. In the earliest clocks a pointer, in the shape of a sun or an arrow or the like, was riveted to the edge of the disk (see Fig. 2); but later on a hand more or less ornate in design was adapted to a square head, projecting from the motion-wheel, and made fast by the knobbed screw which held the disk in place. The disk itself was commonly coated with red lacquer. Such a clock is shown in Fig. 6 (height 14 inches).

Fig. 6.

The clock was surmounted by a large bell seated on a stout iron rod, rising perpendicularly from the top plate of the frame, usually bent over at the top and carrying a screw, which passed through the hole in the bell. The bell was then made firm by a nut. The nut in the early clocks was a feature of Japanese design, being made of wrought iron, and having either two or three wings formed in coils. Beside giving a distinctly ornamental finish, a nut of this form obviates the use of pliers for screwing and unscrewing, as in English lantern clocks. Unfortunately, in later clocks this tasteful terminal was abandoned, and a turned knob was substituted (*see* Fig. 10). Where a common square nut is found to hold the bell, we may be sure that it is a modern replacement.

The vertical bell rod also had one or two arms extended at right angles, grooved at their extremities, to which was tied the cord or cords, used for suspending the balance or balances, as the case might be. A second arm, or pair of arms, projected below, terminating with a pivot hole through which passed the pierced upper end of the verge or verges, the silk suspension cord being threaded through an orifice in the latter.

In order to make the rod perfectly rigid, as was essential for the proper play of the balances, it was made with a broad seating to rest on the top plate, but it was prolonged through a rectangular hole in the plate. In this prolongation was accurately cut a long rectangular slightly tapered eye, through which was passed and driven tightly home a long steel or wooden wedge.

The bell itself was characteristically Japanese in form. In the earliest clocks it was very deep, almost like a bee-hive in shape (*see* Fig. 4). Subsequently it became shallower, and later on its place was taken by a bell almost flat at the top, both in lantern and bracket clocks (*see* Figs. 10 and 11). Both kinds therefore differ from the shallow dome-shaped bells of European make, such as we associate with our own "lantern," long-case and bracket clocks. Another peculiarity of the Japanese bell is that round the inner edge there is a projecting rim. I have been informed, by a well-known

Fig. 7.

London maker of small bells, that English founders are unable to reproduce this feature.

The casing of the old clocks was, as I have said, either devoid of ornamentation, or was but sparsely engraved; but, as time wore on, the fashion for more and more elaborate decoration set in. One form consisted in a sort of "all-over" mottling of the plates (Fig. 7); but as a rule the graving tool was resorted to.

At first the engraving was shallow in broad outline, consisting of sprays of the foliage and flowers of the chrysanthemum, and generally Japanese in character (*see* Figs. 9, 10, 11).

This simpler form of ornamentation would appear to have undergone a change with the introduction of a modification in the casing. The sides were no longer hinged to the back plate, but were fitted, like the front plate, with tongues into slots. A small knob was provided near the top of the plate for pushing the plate into position under the top of the frame or for withdrawing it. So nice is the fit that it holds in position without other fastening (*see* Fig. 8).

It must not be understood that the hinges were entirely abandoned; for we find them on some clocks of the very latest design. The use of independently fixed side plates became, however, much more common. The engraver's art now seems to have fallen under European influence. The charm of native simplicity is lost; and, in its place, we get deeply-cut ornate floral designs, which, towards the close of Japanese clockmaking, degenerated into more or less conventional patterns.

But not only in the decoration do we find the old order changing. The antique cupola-shaped bell gave way to the flat-topped gong, a change which was doubtless found necessary when the clock was covered with a hood. Finally, the old, simple four-square case ceased to please, and turned brass pillars were placed at each corner. For this purpose the bottom and top plates had to be extended all round, to admit of a screw to fasten the pillar at the bottom and a hole for it to pass through at the top (Fig. 8).

I believe that this change took place concurrently with the introduction of a new form of revolving hour circle and of the "bracket" spring-driven clock, in which this form of dial and the corner pillars are almost universal. As these changes mark a new era, and the passing of the primitive Dutch model, it will be more convenient to refer to them when we come to the subject of "bracket" clocks (*see* pp. 247–262).

Fig. 8.

I must, however, mention here that the new form of revolving hour circle and fixed hand offered an advantage over the old system when the clock had only a single balance. As the length of the day and night hours varied with the season, it must have been highly inconvenient and troublesome to have to adjust the position of the little weights on the balance every night after sunset and again before sunrise, in order to arrive at the proper length of the "hours" for each period of the day.

Fig. 9.

In order to overcome this difficulty we find some of the old clocks "converted" to the new style of motion work. The old fixed hour circle with its revolving centre was removed and the new revolving circle and fixed hand with adjustable "hour" plates was substituted.

Fig. 9 (height $9\frac{1}{2}$ inches) is an interesting example of this "improvement," which would seem to have been effected at two different periods. The clock itself is of the old Dutch model, with the casing charmingly engraved with chrysanthemums. It possesses a bell of peculiar shape ornamented with decoration. [The only other embossed bell on a clock in my collection is illustrated in Fig. 23.] The design is in low relief, and represents human figures and, apparently, clouds. The clock had also been fitted with a double calendar, the wheels of which had to be sacrificed to admit of the insertion of the larger new dial in the front plate. At a later date, no doubt, when equinoctial European hours were introduced, and the shifting of the hour plates was no longer necessary, they were set at equal distances and soldered into position at the back. In this way an ancient clock was still made to do duty, although its artistic beauty suffered in consequence. Strange to say, the old iron locking plate is retained.

Fig. 10.

We have seen in a former chapter (p. 136) how our own old lantern clocks were converted into much more accurate timekeepers by the substitution of the anchor escapement and long pendulum in place of the primitive verge and horizontal circular balance. In Japan it was the motion work that underwent alteration, but in some cases the conversion of their antique clocks has been limited to a new twelve-hour locking plate, without change of the original hour circle.

The Japanese "hour," as already explained, was divided into ten *buns*, and these tenths were inscribed on the "hour" circle with a long stroke half-way. I possess a very unusual clock in which these divisions are indicated in a different manner.

Before the introduction of the concentric minute hand, some of the old Dutch clocks indicated the subdivisions of the hour by means of a second small circle, generally placed below the hour circle. This small circle, with its own hand, in the early clocks showed only the four quarters, but towards the beginning of the eighteenth century the

Dutch clocks, made at Zaandam, divided the lower circle into minutes. Clearly, a Dutch clock of this type must have reached Japan and inspired the clock illustrated on p. 227 (Fig. 10).

In this case the small circle is placed *above* the hour circle. The motion of the hand is conveyed from the second wheel of the going train, the arbor of which extends beyond the frame and ends in a square. To this is fitted a large wheel, behind the front plate, geared to another wheel of the same size, whose arbor is prolonged through the front plate and carries the hand. It will be noticed that the dial shows the half-"hour" 半 and the full "hour" 時, each portion being divided into ten parts.

Fig. 11.

Another noteworthy feature of this clock is that the circles are silvered, and the numerals engraved, differing from the more usual practice of lacquering the circle black with its numerals and zodiacal signs in gold.

In other respects, too, this clock, whilst it follows the old Dutch model, gives evidence of much more careful workmanship than those of its class. The engraving of chrysanthemums on the casing is beautifully executed. The wheels are of brass, well cut. The locking plate is also of brass with rounded notches, and is driven by a six-leaved brass pinion attached to a square on the arbor of the second wheel. On the other hand, it has neither alarum nor calendar. (Height 8½ inches to top of bell.)

One other small lantern clock in my collection has a brass revolving hour circle with the numerals 9 to 4 engraved once only.

Another antique clock, which is here reproduced (Fig. 11), possesses unusual interest, in that it gives the name of the maker. It is very strange that the Japanese, who were so particular in signing their works of art, should have omitted to do so in the case of clocks, which in the early days at least must have been regarded as objects of peculiar interest. This is all the more remarkable when it is added that no two clocks are identical, so far as my observation has gone. The explanation may be perhaps that that which appealed to the eye—the casing and its embellishment, the engraved or lacquered hour circle, the gong, and the stand—were all no part of the clockmaker's art, and that he was therefore not entitled to attribute to himself the completed whole.

This view is borne out by the only two clocks in my collection upon which the maker's name and residence are given, one being the lantern clock in this illustration, and the other a bracket clock, of which I shall speak later on.[1]

The inscription in the present instance is engraved upon one of the back uprights of the brass framework. This is interpreted to read as follows :

> Made by Nagamasa Araki Yámato Nojio [of the]
> clan Fujiwara of Yojio Street Kyōto.

The wheels being of brass, and the bell of the shallow flat-top type, would indicate that the clock is not of early

[1] The clock in the Science Museum referred to on p. 220 is also signed. I have bracket clocks with a signature, but without indication of place or date.

date. It will be noticed that the "hour" circle is of most unusual character. The numerals differ from the general form, and have a distinctly archaic appearance, whilst the half-"hours" are represented by round knobs. But here in this clock of old design we have a stationary (copper) hand and a revolving hour-circle; surely a very early example of this innovation, but perhaps a conversion.

It is an unfortunate circumstance that Japanese clocks, with the rarest exceptions, should be unsigned and undated. So far as the absence of date goes, they follow the example of European clocks; but, in regard to the latter, we have records of the European makers and the period when they lived. In treating of Japanese horology, however, we have no guide whatsoever as to the date of their manufacture. At the best we may assume that certain features, such as the material used in the clockwork, the form of the escapement and the "stand," or support, the shape of the bell, the nature of the dial, the design of the case and the character of the ornamentation, indicate progression from the beginning of the seventeenth century down to the abolition of Japanese time measurement in 1873. But even so, in a country where an art or craft was handed down through generations, tradition undoubtedly played a great part; and we should not be justified in concluding, because a clock has all-steel wheel-work, and is made after the old Dutch pattern, that it is either a production of the seventeenth century, or is older than a similar clock with brass wheels. I am therefore averse from any attempt to dogmatize on the subject of age, except so far as to express the opinion that all lantern, bracket and pillar clocks with turned corner pillars were not made prior to some forty years before the crucial date of 1873.

In the frontispiece to this section I have reproduced a colour-print in my possession, which forms one of a series of illustrations of Japanese arts and crafts. It is the representation of the clockmaker exhibiting one of his productions to a lady customer. The date of the series is attributed to about 1830; and we must necessarily suppose that a clock of the period was depicted by the artist. Here

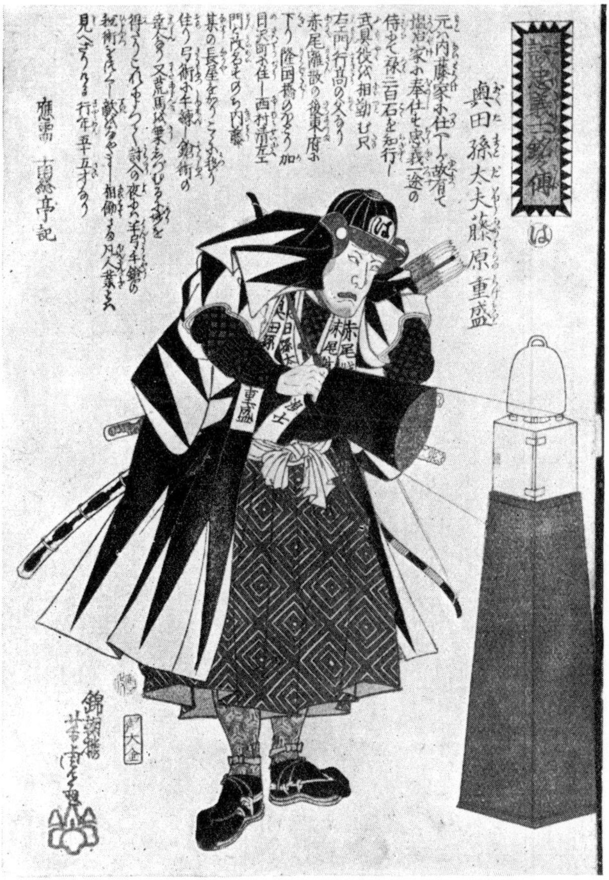

Fig. 12.

then we have a clock of antique design, mounted on a truncated cone, a form of stand associated with the seventeenth century (compare the illustration on p. 216).

I may here mention that clocks very rarely appear in

Japanese prints, which goes to confirm that they were not in common use. The British Museum possesses but one print—a roll print (No. 1418) which represents a lantern

Fig. 13.

clock with a single balance and iron works, and a red revolving centre to the dial. The print is by Shiba Kōkan (1736–1818). He was banished to Nagasaki, where he was in touch with the Dutch traders and wrote on Western art.

Another print in my collection illustrates the story of the Forty Faithful Retainers, one of whom appears to be watching the time for their raid by the aid of his lantern (Fig. 12).

Another cheap colour-print reproduces a clock on a four-legged stand (Fig. 13).

I had it on the authority of Mr. T. Ishiguro, formerly Director of the Art Technical School of Tokyo, when he inspected my collection a few years ago, that the art of *turning in metal* was not practised in Japan before the thirties of last century. Unless this statement can be controverted, we are forced to the conclusion that by far the greater number of the clocks which have found their way abroad, after Japan had discarded them as useless, is of comparatively recent construction.

The English lantern clock with its heavy weights, and the Continental clock of similar character, were attached to a nail in a solid wall, either by means of a loop at the back of the top plate, or they stood on a bracket supported in like manner. This was not always practicable in Japan, for two reasons. Firstly, the solid wall was commonly lacking, and secondly, the custom of sitting on the ground demanded that the clock should be placed at a much lower level. Many of the early Japanese clocks have, it is true, a loop placed vertically, in this respect differing from the English horizontal loop.[1] But the practice was already adopted in the seventeenth century of placing the clock upon a stand of suitable height. Unfortunately, in so many cases the clocks of older design, those which preceded the smaller clock with turned corner pillars, left Japan without their supports ; but I am able to illustrate some of the stands from specimens in my collection.

The earliest form of stand was doubtless that of the truncated pyramid ; but a clock with such a stand as it appears in the frontispiece to this section affords a proof,

[1] The presence of these loops would go to prove that such clocks were in fact made for wealthy owners, whose houses were constructed of solid masonry.

from the date of the print, that this form of support was continued well into the last century.[1]

An excellent example is the clock in the Science Museum, to which reference has been made. Here the wood is covered with dark brown lacquer, with some applied metal work. It may be attributed to the end of the seventeenth century, being of the same type as Fig. 5.

A far more elaborate stand in my collection, 39 inches in height, is shown in the illustration on p. 216. It is decorated with Somada lacquer, enriched with beautiful designs of flowers on three sides in mother-of-pearl, and having on the front panel—which is removable for drawing up the weights—an inscription also in mother-of-pearl. Fig. 14 shows two of the other panels.

Fig. 14.

The interpretation of the legend caused much perplexity, but it was finally deciphered by a professor at Tokyo. It appears that it is written in archaic *Chinese* cursive script, whence the difficulty in reading it. The meaning appears

[1] A clock surmounting a low pyramid stand is exhibited in the vestibule of the Guildhall Museum, London. The mechanism suggests that it belongs to the latest period. The bell is struck by a "Jack."

to be that "in place of a suspended pot this present instrument will give the hour and division, correctly representing Heaven's word," i.e. the astronomical observation.

According to one interpretation the "suspended pot," or

Fig. 15.

Fig. 16.

jar, was a form of clepsydra ; according to another, a hollow vessel which was struck as a time signal in the temple. The stand is attributed to the latter part of the seventeenth century. It has been used to support the Tokugawa clock, illustrated on p. 219.

Another form of stand was a support with four straight or cabriole legs, a type which continued to be associated with "lantern" clocks, that may be attributed to the eighteenth century, and which, in smaller and more delicate design, to suit the diminished size of the clocks with turned corner pillars, was retained until the end of the Japanese period.

Fig. 17.

Fig. 15 gives an illustration of the older type of such a stand which is embellished with a certain amount of coloured design in lacquer.

With the advent of the clocks of the latest period of native clockmaking, we get stands of plain woods, stands with carving, and lacquered stands. Examples of these different styles are here illustrated.

Fig. 16 illustrates a stand in carved wood in the form of trees decorated with a certain amount of gold lacquer. Height overall, $25\frac{1}{2}$ inches.

Fig. 17 reproduces a very beautiful specimen in black and gold lacquer containing an exquisitely finished clock (Fig. 8). The dial and corner pillars are silver, and the side plates are engraved inside as well as outside. The pin on the right of the dial can be inserted in any of the holes, and at that hour the lifting piece on the left is raised to discharge the alarum. (Height overall, $29\frac{1}{2}$ inches.)

The most striking specimen in my collection of the latest

Fig. 18.

period is shown in Fig. 18. Here the stand and the hood are of blue and white porcelain. All four sides of the clock are of beautifully pierced and engraved fretwork. The dial is gilt. It has a circular balance. (Height overall, 28 inches.)

Another blue and white porcelain stand is shown in Fig. 19, but in this case the hood is unfortunately missing.

Fig. 19.

A departure from the two prevailing models of stands is illustrated in Fig. 20. In this case the cords and weights are concealed in a tall rectangular case, with lattice-work panels, the whole embellished with lacquer of many colours. It is altogether lacking in the repose and dignity associated with Japanese art. Judging from its height (48 inches), it was probably a temple clock.

Less frequently, clocks of the lantern type, both of older and later work, were enclosed in hooded brackets, and here again we may suspect Dutch influence. Figs. 21 and 22 illustrate clocks of this description. The former is of the older type with bar balance and plain case; the latter has the corner pillars and a circular balance.

Fig. 20.

So far we have been dealing with the external aspects of the lantern or weight clock, and we have now to consider the more important portion of the subject—the clockwork proper.

At the beginning of this chapter I have to some extent anticipated a description of the mechanism in referring to the balance and the locking plate. The point that I wished to emphasize was that the Japanese clock in its earliest form was, so far as the wheel-work is concerned, a mere replica of the European weight clock of the period. In previous chapters I have described the operation of the mediæval clock, and I will only repeat here that the general design underwent but little modification between the beginning of the fourteenth century and the introduction of the pendulum as the governor in 1657.

Fig. 21.

So far as the "going" train, or series of wheels terminating with the escapement was concerned, the mechanism consisted merely of a great wheel, carrying the cord and driving weight, a second wheel driven by a pinion on the arbor of the great wheel, and lastly a vertical "crown"-wheel, driven off the second wheel.

Fig. 22.

This simple train was employed in the smaller clocks of the early type, whereas in the larger specimens a third wheel generally drives the 'scape-wheel.

I have already explained the function of the balance as the means of controlling the speed of the train; but in

Japan the swing of the balance itself had to be accelerated or retarded to correspond with the length of the day and night " hours," and to adjust the speed in harmony with the particular period of the year.

For this purpose the serrated arms of the balance, or of both balances when two were used, were divided by vertical lines at the end of each group of five notches.

The method of regulating a *double* balance clock is thus explained by Siebold in his monumental work on Japan.[1] At the summer solstice, the weights on the balance controlling the day " hours " were suspended from the outermost notch, whilst those on the night balance were hung on the innermost notch. At the next period of adjustment of the duration of the respective " hours " the weights on the day balance were moved one notch inwards and those on the night balance one notch outwards, with the result that at the equinoxes all the weights were in the same position. He states, however, that each balance had thirty notches on each side, the movement of one notch being effected at the end of 6 d. 4 h. 12 m. This latter statement is inadmissible. Not only had the Japanese no means of precise time measurement, but no one would make adjustments at such odd intervals. The change was doubtless effected at the commencement of each half-month. Moreover, as we have seen, the day and night " hours " were not of equal length at the equinoxes. Furthermore, from an examination of all the balance clocks of archaic type in my possession, I find that some have thirty and others thirty-five notches ; and this variation occurs in specimens which appear to be of early date. Some even have a few more notches than thirty-five.

Doubtless, the number of the notches was determined by the number of vibrations that the particular train of wheels would cause to the balance. What that number was to be appears to have followed no rule. I have counted the trains

[1] " Nippon," by Phil. Franz von Siebold, Leyden, 1852.

of all the clocks with steel wheels in my collection, and the variation in the number of beats to one turn of the great wheel is so great that no conclusion is possible. I believe that with European balance clocks there was also no fixed rule in the determination of the numbers of the teeth in the wheels and pinions. This was due, no doubt, to each clock being an individual production.

To illustrate the divergence, these are the going trains of two very early clocks :

Great wheel . .	72—6	Great wheel . .	66—10
Second wheel . .	72—6	Second wheel . .	66— 7
Crown wheel . .	25	Third wheel . .	54— 7
		Crown wheel . .	15

At what period did the Japanese clockmaker hit upon the ingenious plan of a double escapement with one balance for the day "hours" and the other for the night "hours"? The inconvenience of having to set the weights on a single balance twice a day at a certain time must have been very great ; nevertheless, the single balance remained in use in lantern clocks of what I will call the archaic type—that is to say, clocks other than those of the latest design with the turned corner pillars—until well into the nineteenth century. They are so illustrated in the prints reproduced in Figs. 12 and 13.

A beautiful and perfect specimen of the double-balance clock is illustrated in Fig. 23.

I must here endeavour to explain the action of the double balance. Simple as it is in operation, as are so many clever devices, it is by no means easy to describe in writing. The change, as we have seen, between the two divisions of the day took place at the stroke of six. A projecting pin was fixed at that point in the locking plate at the back of the movement. Necessarily there were two suspended verges, the longer carrying the upper (day) balance and the shorter the lower (night) balance. These verges with their pallets engage alternately in two crown-wheels, set back to back on the same arbor. Action takes place in the following way: a rod is pivoted at one end in the back frame and at the other end in a bracket. Slightly above this are two long narrow steel or brass tongues, each pivoted at the side of the frame. These tongues at their ends have each a shallow groove in which the end of the respective verges can rest. At the extremity of the rod is fixed a four-leaved cam, or alternatively a star-wheel, in which the pin on the locking plate engages. The rod is pierced at points below the tongues, and pins are driven through it, projecting at right angles to one another on each side of it. (In some clocks a steel triangle takes the place of the pins.) What happens then is this. The locking plate is set free to revolve when the striking train is set in motion at the " hour " of six. Its pin then engages in the cams, or star-wheel, which accordingly causes the rod to turn. Thereupon one of the two pins below the tongues rises from the horizontal to the vertical, and in so doing pushes up its tongue, which in turn raises its corresponding verge to such a height that the pallets no longer engage in its crown-wheel. Simultaneously the other pin recedes from the vertical to the horizontal, and thus lowers the second tongue into the position in which its verge and

Fig. 23.

pallets can bring its balance into action. At the succeeding stroke of six, the action is reversed, and so the alternate swing of the balances proceeds automatically.

The clock also illustrates an old European method of discharging the alarum, which may also be seen in Figs. 7 and 15, and in "bracket" clocks, Figs. 24 and 25.

The revolving disk which carries the hand is furnished with holes, against which are inscribed the numerals of the "hours" at which it is desired to make the alarum sound. A pin, projecting through the disk, is then inserted in the particular hole, which at the given time raises the alarum lifting piece. The weight is thereupon set free to drive the train, causing the hammer at the top of the verge to strike the inside of the bell with alternating blows.

Alarum work was frequently employed in the old lantern clocks; but unfortunately it has often been removed. The earliest instance of it in my collection occurs in the clock dated 1695, referred to on p. 220; but in this case the discharge is effected from inside the frame.

That the single balance should have persisted is all the more strange, because the escapement with the duplicated foliot balance lent itself peculiarly to the Japanese system of variable "hours." If the first clocks introduced into Japan had been such as had an escapement with a bob or a long pendulum the problem of automatic change in the length of the day and night "hours" would have been one of far more difficult solution.[1]

The earliest clock with a double balance that has come to my notice is that in the Science Museum, to which allusion has been made on p. 220. As the general appearance of this clock agrees with that illustrated in Fig. 5 and the clock of similar damascened casing referred to on the same page, it is fair to assume the end of the seventeenth century as the date of the Science Museum clock.

[1] In the Introductory chapter (p. 194) I have mentioned that Hosokawa's book in 1796 only deals with the construction of a single balance clock.

The striking mechanism follows that of the Dutch prototype, the pins for raising the tail of the bell hammer being sometimes on the great wheel and sometimes on the second wheel. The locking plate, which governs the number of the strokes, is so notched as to strike a single blow at the half-" hours " in the early clocks. The absurdity of this system, which prevails to the present day in European clocks, was not so apparent in Japan, where no " *hour* " struck one ; but in course of time the Japanese clockmaker perceived that, with the long duration of their " hours " at certain periods, it was desirable to give some better indication of the particular half-" hour." In order therefore to make a distinction, the locking plate was so cut that the clock struck *one* blow after the uneven " hours," 9, 7 and 5, and *two* blows after the even " hours," 8, 6 and 4.

When this improvement was introduced I am unable to say, but I find that in my collection it is associated with one double balance *archaic* clock, but not on any other specimen, whilst in those with a single balance the single blow at every half-" hour " is retained.

In clocks of the later corner pillar type the two blows after the even " hours " were regularly employed.

Another problem is the period when the date-indicator was added to the mechanism. The earliest clocks did not possess this addition, as is only natural, seeing that the European clock did not introduce it before about 1680 as a general practice. Perhaps we may here find some clue as to the antiquity of a Japanese clock. It does not follow, because a clock has no calendar, that it is of early date ; but I think that we may assume that the presence of the calendar indicates that it would not be older than the close of the seventeenth century.

As I have already stated, a clock may be furnished with either a single calendar showing consecutively the progression of the twelve signs of the zodiac, or a double calendar giving the combinations of the zodiac with the ten elements and sub-elements, thus completing a cycle of sixty days, according to the table on p. 208.

Out of the fourteen single-balance weight-driven clocks of the archaic or older types in my collection only two possess a date-indicator. The first has the all-steel wheelwork and the deep beehive bell, associated with early work, and it strikes one blow only at the half-" hours." This would therefore appear to represent an early instance of this addition (Fig. 7). The second, already described and illustrated in Fig. 9, is of later date. It has steel works, but is fitted into a brass framework, and it strikes two blows at the even half-" hours." Moreover, the sides of the case are not hinged, but are of the push-in type, to which allusion has been made on p. 224. This clock originally had a double calendar.

From all that has been said, we must, I submit, come to the conclusion that no hard-and-fast rule can be established for estimating the antiquity of any given lantern clock, prior to the introduction of what may be termed the standard type of small clocks, fitted with turned corner pillars and the revolving hour-circle, described on p. 225. Some general inferences may, however, be drawn from the character of the workmanship, always bearing in mind that tradition may have exercised an important influence on each individual craftsman.

To summarize, it is evident that clocks entirely composed of crudely cut steel wheels and steel framework would represent the first efforts in imitating the old European model. The bells would be of the deep beehive type, and the cases generally of copper, either plain, silvered or sparsely engraved. There would be a single balance, and only one blow would be struck at the half-" hours." The inner disk would revolve, having the " hour "-index attached to it. Towards the close of the seventeenth century the double balance would appear to have come into use, though not of general adoption, as also the improvement of the double blow at the even half-" hours." A concentric hand also makes its appearance, and an alarum would come into use.

The next step forward in evolution would consist in the gradual introduction of brass, both in the casing and in some minor parts of the mechanism, such as the fly. The cases would be mottled, or engraved with restrained designs of purely Japanese character, generally chrysanthemums. The bell now becomes shallower, and the calendar is added.

The following stage inaugurates the adoption of brass instead of steel wheels. The engraving of the cases is losing in delicacy under European influence, and the bell is shallow or quite flat.

The final development is characterized by the great departure in design, consequent upon the addition of the turned corner pillars, which have been described. Together with this change the double calendar became general, though not universal, whilst the stand enclosing the hooded clock was reduced to a little over 2 feet in height, a size convenient for a people accustomed to sit on the floor. The clock itself is proportionately small, and its engraving is of coarse and stereotyped pattern. In most clocks of this type the circular balance now takes the place of the old bar balance. It must not be assumed, however, that this new departure in the governing element and the adoption of the smaller clock and stand caused the abandonment of the older model, which may well have continued to be made until all Japanese clocks became obsolete for purposes of time measurement.

In conclusion, it must be mentioned that these clocks require winding twice a day or even more often, resembling in that respect the English lantern clock before the advent of the pendulum.

"Bracket" Clocks

Leaving clocks of which the motive power was a falling weight, we must now pass on to consider the clocks which I include in Class 2, namely, those driven by a coiled spring. Their general design was no doubt inspired by the spring clocks of Europe, which, as their designation denotes, originally rested on a bracket ; although nowadays they usually occupy a position on a mantelpiece or a cabinet.

That which the Japanese spring clock has in common with the European prototype is that the clock is enclosed in a wooden case, glazed on every side. But there the resemblance ends ; for the cases are of much lighter construction, held together only with glue, and they are all of comparatively small size. The largest clock-case in my collection measures 13¾ inches overall (Fig. 38), and the smallest 4 inches (Fig. 33) ; but the general height is from 8 to 9 inches. The cases are made of Shitân, persimmon and other woods, usually without decoration, though occasionally lacquer and inlay were used.

Seeing that the spring-driven bracket clock was in common use in Europe during the last quarter of the seventeenth century, it is remarkable that about 150 years would seem to have elapsed before its general adoption in Japan. I do not consider that any of these clocks in my collection can have been made prior to 1830. They all, more or less, apart from the motive power, possess as to finish the same general characteristics as the lantern clocks of the latest period. The bar balance for regulating the escapement was generally discarded in favour of a circular balance controlled by a spiral "hair-spring," with which, on a smaller scale, we are familiar in our watches.

There was, however, apparently a transitional stage in which most of the features of the old lantern clock were retained. I possess two "bracket" clocks, both of which have the double bar balance and the stationary "hour" circle, with the revolving disk carrying the hand and pierced for the alarum-pin (Figs. 24 and 25). Furthermore, their locking plate and its detent are *inside* the brass case, corresponding with the old style lantern clock, therein differing entirely from the striking mechanism, which,

Fig. 24.

with the new motion work and other changes, constitutes such a break with old tradition. On the other hand, both of the clocks are furnished with the turned corner pillars, and their brass casing is covered with the coarse and decadent engraving with which that innovation is associated.

Although the use of the spring motor would not appear to have been generally adopted before the second quarter of the nineteenth century, the possibility of experiments in that direction must not be excluded. I have, in fact, seen in the cabinet of a Paris collector an elaborately lacquered clock, enclosed in a lacquered travelling case, which has a spring movement, and which, from the character of its

Fig. 25.

wheelwork, suggests that it is of older date than those which we are now considering. This, however, is the solitary instance that has come to my notice.

When speaking of the latest development of the lantern clock, I said (p. 225) that this would be a convenient place to refer to it more particularly. The reason is that I incline to the belief that the introduction of the spring clock brought about the modification of the weight-driven clock, and not vice versa.

It is hardly too much to say that the new design constitutes a revolution in the horological art of Japan and establishes a new era. We pass at once from the archaic, carrying us back to Europe of the fourteenth century, to the modern influences of the nineteenth. As the old Dutch weight clock served as a model for the early Japanese " lantern " clock, so it would seem that the new spring clock was inspired by the Swiss travelling clock. The carefully finished wheels and pinions bear a striking resemblance, and the circular balance with its hair-spring may well owe its origin to the same source.

Nevertheless, the Japanese method of time measurement, with its day and night " hours " of varying length, demanded an hour circle, differing altogether from that which served for the equally divided twelve-hour circle of Europe. The problem was solved by the invention of the rotating circle, shown in Fig. 9, and in the dial of the clock illustrated in Fig. 26.

As will be seen, there is a sunk ring between the outer rim and the solid centre disk. This space between the two serves to receive small plates bearing the numerals corresponding with the six " hours " twice over. Between the " hour "-plates is a series of plain plates of half the width, which indicate the half-" hours." In order to hold the plates in position a circular slot is cut through the sunk ring. Fixed to the back of each plate at right angles is a projecting piece, the base of which has an oblong section, so that it fits nicely into the slot, without being too tight. This is drilled, so that when the plate is placed in position a slightly sprung oblong metal washer, made fast by a pin passed through the hole, makes the plate secure. The plates are not so tight, however, as to be immovable. By means of the fingers or with fine pliers their position can be shifted in the sunk ring, and in this way the adjustment of the varying length of the "hours" is effected. One set of plates represents the day "hours," and the other the night "hours." The plates are generally silvered, but in some clocks the night "hours" are made black.

Fig. 26.

It follows that, when the light period is at its longest, the day plates are spaced far apart, whilst the night plates are drawn closely together, the reverse adjustment taking place as the day-"hours" shorten. The adjustment, which would be made at each half-month, must have been one of great nicety ; and it is difficult to imagine how it was accomplished with any exactitude, especially when the dial was very small. As a guide, the outer rim is graduated with fine lines into ten parts, each part being again subdivided into tenths. The inner disk is also graduated into twelve divisions and half-divisions.

The piece projecting from the hour plates did not merely serve to hold the plate in position in the slot. It answered a double purpose. It was made of such length that, as the circle revolved, each pin in turn raised the lifting-piece, behind the front plate of the clock, which sets the striking work in action. Thus, whether the plates were nearer or farther apart, each "hour" and half-"hour" was correctly struck in correspondence with its numeral and the season of the year.

This new striking mechanism brought about a modification of the locking plate. Instead of the rough wheel of the old "lantern" clock with its square notches, we get a small wheel with serrated teeth of fine workmanship. The lifting-piece behind the "hour" circle is attached to an arbor, passing through to the back plate, and having at its other end another lifting-piece. This latter has a returned end which holds in position a "gathering pallet." As both lifting-pieces are raised together, and the striking train is unlocked, the gathering pallet, attached to a wheel of the train, is free to revolve with it, and engages in the teeth of the locking plate one after another, thrusting the plate round until the proper number of blows has been struck. At that moment a piece, projecting from and forming part of the back lifting-piece, drops into the corresponding deep tooth of the locking plate, and brings the train to a standstill. The device is shown in Fig. 34.

This device of the gathering pallet was no doubt an adaptation from the rack striking work, invented by Edward Barlow, which generally superseded the old locking plate system in English clocks from the beginning of the eighteenth century (see Fig. 29).

Recurring once more to the latest design of the " lantern " clock, I must here emphasize that it did not necessarily incorporate all the changes of mechanism introduced with the spring-driven clock, retaining only the motive power of the falling weight. On the contrary, the old form of dial, the double bar balance and the old system of striking were

Fig. 27.

frequently retained. The similarity consists rather in the same finer wheelwork, the same style of engraving, the adoption in many cases of the circular balance with the revolving hour circle, and the introduction of the corner pillars, these latter developments probably representing the final phase of the evolution of the " lantern " clock.

Returning now to the " bracket " clock, we must suppose, as I have said, that the refinements of its workmanship were gradual. I illustrate from my collection (Fig. 27) what is perhaps one of the earliest specimens of the new type with the new style " hour " circle and locking plate, the circular balance, and with the corner pillars.[1] The wheelwork is still entirely of steel, with exception of the brass barrels and fusee and the fly. The engraving of the front plate is coarse and the back plate is plain. But its chief interest consists

Fig. 28.

Fig. 29.

in the inscription engraved on the inside of the back plate, which has been translated as follows :

" Made by Kōki Kobayashi living in the province of Settsu in the 8th month of the second year Tempo period " = 1830. As Osaka is in the Settsu province, the clockmaker probably resided in that important city.

I have mentioned the rarity of names and dates on Japanese clocks, and I therefore think it reasonable to assume that, in recording his name and residence with the

[1] The front slide has been removed for the illustration.

year and month of his work, the clockmaker desired to signalize a notable achievement. The date corresponds with that assigned to the earliest use of metal turning. The train moreover is constructed to run for about five days.

I have mentioned the rack-striking mechanism. It is strange that the Japanese clockmakers, when they were adopting new methods and designs, did not make general use of this system, which possesses the great advantage that it ensures the proper number of blows in accordance with the position of the hour hand, however it may be placed; whereas, with the locking-plate, if the strokes are not in correspondence, it is necessary to make the clock strike by raising the lifting-piece until the two are once more in agreement.

I own, however, a clock which possesses the rack; an exquisite specimen, 11 inches in height overall. The front plate is of finely chased fretwork, embellished with artificial gems, the back plate of chiselled and engraved fretwork. A compartment below the movement contains a musical box, set in motion by the clockwork. The tune is not European, but the musical box suggests Swiss manufacture or an imitation. The surround of the circular balance is of chased silver, and the finely turned corner pillars are gilt (Figs. 28 and 29).

As a rule the movement of the "bracket" clocks is set between oblong front and back plates, either solid or of fretwork with chasing of floral designs; but occasionally we meet with dials of circular form. Fig. 30 illustrates a very charming piece of this type,[1] the front plate of which has already been given in Fig. 26, to show the revolving "hour" circle. The height of the case overall is 10½ inches; and it is the only clock in my collection which possesses the ring, revolving above the "hour" circle, which gives the twenty-four seasons (see p. 213). The front plate and the corner pillars are gilt.

[1] Front slide removed for the illustration.

I possess another clock with circular plates ; but in this case the front plate is covered with opaque turquoise blue glass, and the movement is surrounded at the base by a gallery of engraved fretwork.

Fig. 30.

The next illustration (Fig. 31) represents another clock (front slide removed), having a musical box in the base. Apart from this, it is a departure from the usual design, inasmuch as the going train is controlled by a verge escapement and short pendulum, in place of the usual horizontal circular balance, therein resembling Fig. 39. Another interesting feature of this clock is that the front plate is decorated with cloisonné enamel. Seeing that this work was not practised in Japan before 1838,[1] we have a further indication that "bracket" clocks of this type are of comparatively modern origin.

The next illustration (Fig. 32) is that of a remarkable and beautiful clock which calls for special description, combining as it does characteristics of both the weight-

Fig. 31.

driven and spring-driven models. Externally, we get the turned corner pillars, the engraved side plates, which are lacking in the glazed wooden cases, and the new revolving "hour" circle. But there the resemblance to the usual type of bracket clock ends. The wheelwork, apart from the brass barrels, fusees and fly, is entirely of steel, and the striking mechanism is on the old Dutch system, with a brass

[1] "Encyclopædia Britannica," 11th edition, vol. 15, p. 189.

square-notched locking plate outside the back plate. The mechanism possesses some unusual features. The fusees, instead of being placed above the barrels, as in English bracket clocks, are planted on the same level as the barrels at opposite ends of the frame. This arrangement is necessi-

Fig. 32.

tated by the Japanese preference in their " bracket " clocks for a frame greater in length than in height. The same arrangement occurs in the clocks with the double foliot balance (see Fig. 25).

The going train is wound by two equal sprocket-wheels, the lower one being made fast to the fusee. The movement is governed by a circular balance and hair-spring, with the sliding regulating weights on its arms ; but this shows signs of having been substituted for an original foliot balance. The clock runs for about ten days. The front plate is embellished with exquisitely chiselled and chased chrysanthemum flowers and engraved foliage. The back plate

Fig. 33.

has a similar design in flat engraving. There is a double calendar. A portion of the spring alarum is unfortunately missing.

The movement is enclosed in a case, to the beauty of which no illustration can do justice. The wood is light brown, with such a wonderful figure that at first sight it looks as if the whole case were covered with brown lacquer. The decoration of this surface is entirely in gold lacquer in relief.

Fig. 34.

On one side is a large pine tree with huts ; on the other, below, cranes standing in reeds, and above, a hill crowned with huts and a smaller pine, with a crane in flight. A small tree, a crane and reeds occupy the small space of the glazed

Fig. 35.

Fig. 36.

front, and the top is also ornamented with cranes. From this description it will be seen that the clock is in all respects a product of purely Japanese workmanship and art.

The next "bracket" clock that I illustrate is the tiny specimen referred to on p. 247. It is a little gem of exquisite workmanship. The clock itself measures only $2\frac{1}{4}$ inches in height. The front plate is of delicate fretwork beautifully engraved, laid on a plate of blackened steel. In other respects it is a miniature edition of the standard type. It is enclosed in an ebony case glazed on all sides, resting on an ebony stand of the usual Japanese pattern (Fig. 33).

Fig. 37.

Although the turned corner pillars are generally associated with clocks of the bracket type, their use was not universal. Fig. 34 shows a neat little clock in a beautifully figured "streaky" persimmon case, only $4\frac{3}{4}$ inches high. The plates are quite plain. It has a double calendar. The back illustrates the new locking plate.

Another plain little clock has a verge escapement, with a bob pendulum swinging in front of the dial and only one series of numerals (Fig. 35). Side by side with it is the movement of a small clock with a circular front plate, and the spring balance set vertically (Fig. 36).

The next illustration (Fig. 37) is that of a dainty little upright clock with a single foliot balance, spring-driven, with mechanism of the old lantern pattern. The brass casing is entirely covered with landscape design in very low relief. The outer wooden case is so plain as to suggest that it was only intended as a protection from dust, and for travelling.

I conclude the section on Bracket Clocks with a description of the large clock, mentioned on p. 247, the front and back of which are here illustrated (Figs. 38 and 39).

This beautiful clock is fitted with a musical barrel placed above the movement ; but unfortunately it has been subjected to severe " conversion " by alteration of the motion work so as to carry a " minute hand." This was, doubtless,

Fig. 38.

Fig. 39.

done when European time was introduced, so that the clock might still be serviceable. In order to effect this change the lunation work has been put out of action, and the circle which showed the twenty-four seasons (see Fig. 26) has been sacrificed. It has, moreover, the unusual feature of a verge escapement with bob pendulum. The case measures 13$\frac{3}{4}$ inches in height.

" Pillar " Clocks

We now come to the third of the three categories of clocks of typical Japanese design. As I have remarked, this form of clock has no exact counterpart of European origin, although some resemblance is to be found in Martinelli's water-clock, which marks the hours on a scale by means of a descending index.[1]

Clocks, as I mentioned in my introductory chapter, were not in common use in Japan, their cost no doubt restricting them to the well-to-do classes. Poorer people had to depend on the clock in the temple and on the position of the sun for estimating the time of day. But the obvious convenience of possessing a more or less accurate time measurer must have created a demand for something less expensive than the lantern or bracket clock. The clocks which we now have to consider were doubtless devised to meet this demand. They required no stand or piece of furniture for their support, which in most apartments would be regarded as an obstruction. A nail, driven into some pillar or other permanent portion of the structure of the dwelling, was all that was necessary for their installation.

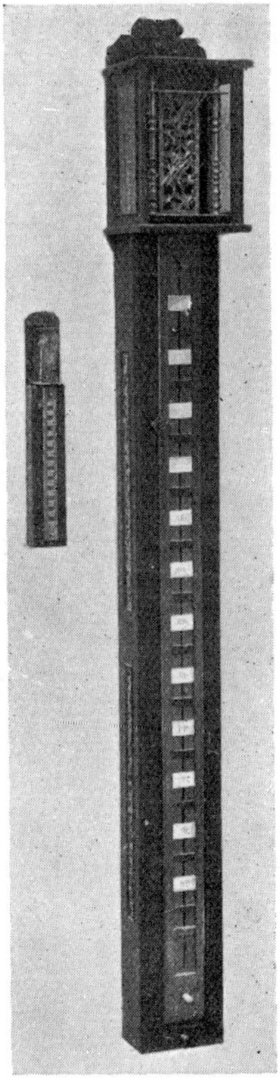

Fig. 40.

The clock consists merely of a long, narrow, wooden case, on the top of which is planted the going train enclosed in a glazed hood, suggesting somewhat an English long-case clock in miniature. The length of the case varies considerably. The longest in my possession measures 47 inches and the shortest $10\frac{1}{2}$ inches, but I have seen a clock even smaller than the latter. The illustration (Fig. 40) will serve to show the two extremes.

[1] A working model may be seen in the Science Museum, South Kensington.

The movement, much greater in height than in width, is commonly surrounded by four turned corner pillars which hold together the top and bottom plates ; but in what appears to be the earliest form of these clocks the pillars are wanting. The clock is held in position by two screws, which pass from the back of the case into two holes in the back plate. Fig. 41 shows the ordinary type of movement, the front plate in many cases lending itself to much more ornate treatment, as reference to later illustrations will demonstrate ($4\frac{1}{2}''$ overall).

Fig. 41.

Where all is conjecture it is only possible to suggest the evolution of this type of clock. In some cases the front plate is of solid brass, with or without engraving and without corner pillars ; whereas the generality of them has front plates of fretwork more or less elaborately engraved, and well-turned pillars. The more finished style of the latter may be taken to indicate progressive design. The illustration (Fig. 42) shows the very beautiful front-plate of the clock reproduced in Fig. 51.

The going train is governed by either a bar balance (see Fig. 43), a circular balance, or by a verge escapement with a short bob pendulum swinging in front. Judging from the specimens in my collection and from many others that have come under my notice, I should say that the circular balance was by far the most commonly adopted.

Of those in my possession two have the bar balance, six the bob pendulum and seventeen the circular balance.

One of the first category is of very primitive construction, representing, perhaps, a first attempt at the production of a cheap clock. It consists merely of an oblong slab of black lacquered wood, to which is screwed a brass "hour" scale,

Fig. 42.

the time being read from the position of the top of the descending weight. The movement is enclosed in a glazed hood. There being no drawer to contain a key, a winding handle is attached to the great wheel.

The second clock with the bar balance is cased; but the wood is of poor quality, and it has the unusual feature that the "hour" plates are all engraved with the signs of the zodiac, instead of the customary numerals. There are no plates to mark the half-" hours," no corner pillars, and the front plate of the clock is solid (Fig. 43, height 16 inches).

The simple form of these two clocks, taken in conjunction

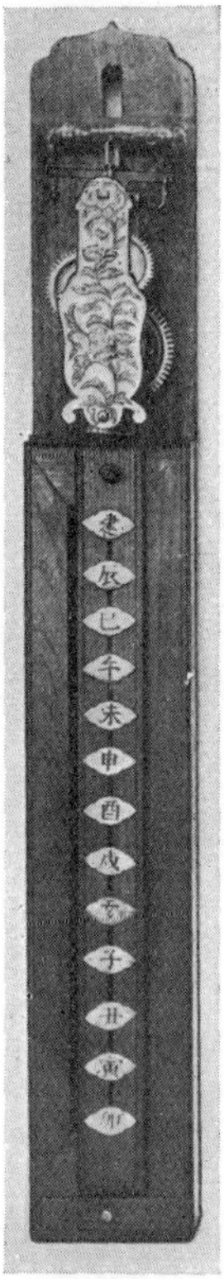

Fig. 43.

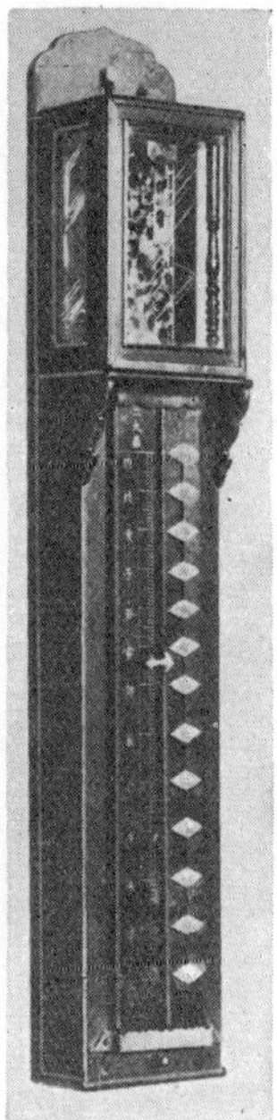

Fig. 44.

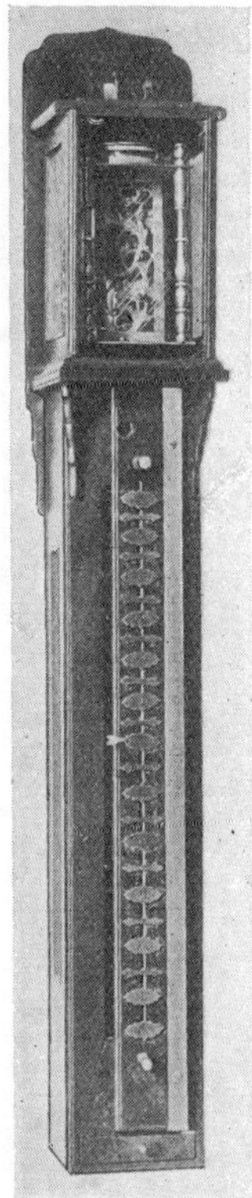

Fig. 45.

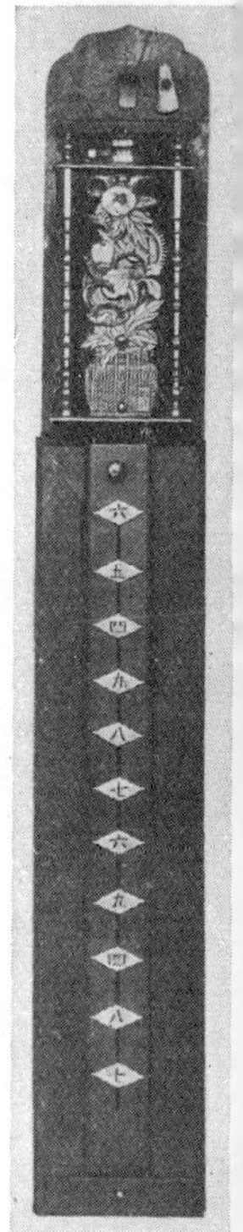

Fig. 46.

with the foliot balance, would seem to point to the first stage in the evolution of this new type.

Fig. 44 illustrates a pendulum clock (height 19¾ inches), and reference may be made to Figs. 1, 42 and 53 a.

There are many variants in these pillar clocks, ranging from the plainest to the most highly finished productions.

Even such details as the "hour" and half-"hour" plates do not appear to have been made to a standard pattern. The principle is, however, the same in all the various types. To the body of the case below the hood is attached either by a screw or screws with a long turned head, or by a sort of small turnbuckle, an oblong slab with a groove running down its centre for inserting the "hour" plates. The aperture in the case is rather wider than this slab, so that, with the slab in position, a space is left at one side of the slab, down which the index or pointer is able to travel.[1] The pointer, being fixed to the leaden weight which drives the going train, descends with the weight, and thus indicates the time as it passes the "hour" plates in succession. As in the case of the bracket clocks, the plates are furnished at the back with a projecting piece of oblong section pierced with a small hole. This is passed through the groove or slot in the slab, and is held in position by a slightly convex brass washer having a corresponding oblong aperture and then fastened by a pin through the small hole. The plate is not made too tight, however, so that it can be moved up or down in the groove; and in this way the plates are drawn closer together, or extended farther apart in accordance with the lengthening and shortening of the "day" and "night" hours respectively. Such adjustment would take place bi-monthly. In order to facilitate

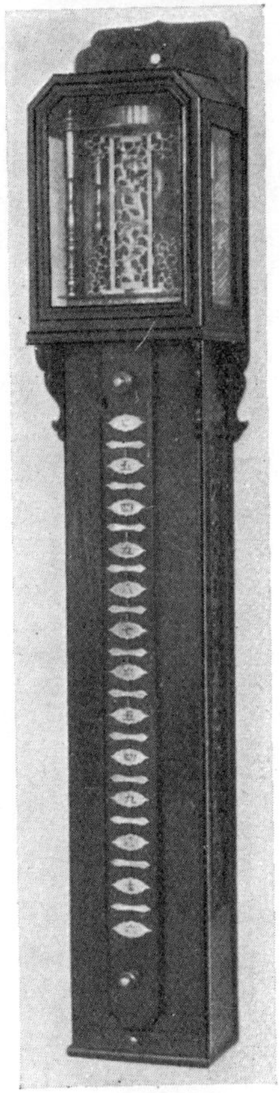

Fig. 47.

[1] Unfortunately, the pointer and some of the "hour" plates are often missing.

the accurate spacing of the numerals, some clocks have a brass plate fixed at the side of the "hour" slab. It would seem to be a later addition, made when the clock was used to show equal hours. Such a clock is shown in Fig. 45 (height $20\frac{1}{2}$ inches).

Fig. 48.

The simplest clocks are provided with "hour" plates only, whilst those of more elaborate design have narrow unnumbered plates between them to indicate the half-"hours."

To illustrate the former type I give a delightful little timepiece, the front plate of which represents a basket of flowers, beautifully chiselled and engraved, bearing in a cartouche the artist's signature Kuniyuki—a rare occurrence (Fig. 46, height $15\frac{1}{4}$ inches).

A very fine striking-clock with the half-"hour" plates (hand missing) is shown in Fig. 47.

One very interesting specimen demands special mention. It would appear to have been made for an astronomer, because the "hour" plates, instead of the ordinary numerals, are engraved with constellations, the number of the stars corresponding with the respective "hours." Moreover, the top plate of the movement has the Great Bear engraved on one face and another constellation on the other. But the most remarkable feature of this clock is that it has a bell at the top of the movement, which by an ingenious contrivance is rung once a day by an alarum (front slide removed). The discharge is effected by means of a long lever enclosed in a compartment on the left side of the case and terminating in a cam. The "hour" pointer in descending passes over this cam, and so thrusts back the lever, which then at its other end unlocks the alarum movement, and so sets the bell ringing. The object of this signal was presumably to warn the astronomer that the moment was approaching for taking an observation of the sun. The circular balance is set *vertically* outside the front plate, its usual position being occupied by the bell (Fig. 48). It is also a striking-clock (height 19¾ inches).

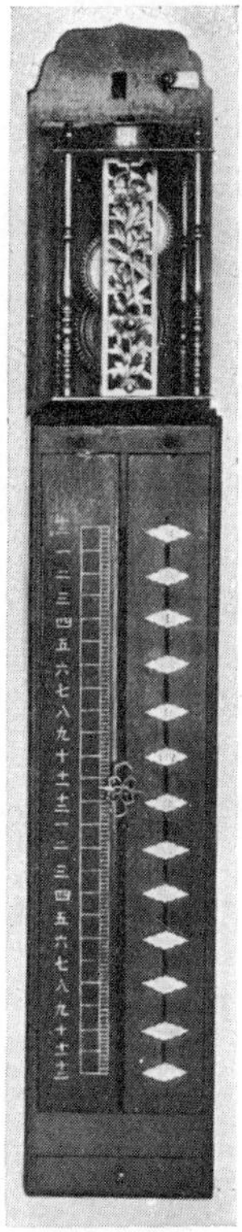

Fig. 49.

It is obvious that the adjustment of the plates to correspond with the duration of the "hours" according to the time of the year must have presented a very difficult problem for the class of people who would possess such a cheap form of clock, the assistance of an expert being probably required. To obviate this difficulty a new device was invented.

Whilst in most cases the slab with its "hour" plates was retained, there was placed on the other side of the aperture, down which the pointer descends, a black-lacquered scale, graduated in gold, with the "hours," divided also into tenths, and correctly divided in accordance with the particular half-month. This is well illustrated in Fig. 1. In order to provide for the whole year a set of seven such scales was supplied in a box. At the top of each scale is inscribed the month at its commencement or the half-month to which the scale was applicable. For instance, if a plate bears the Japanese numbers for 2—8—½, it would mean that, according to our reckoning, the length of the "hours" would be the same in the second half of March and September, but, of course, in inverse order; so that the "hours" commencing with six at the top of the scale down to the following six would represent the night "hours" in March; whereas they would be read as the day hours in September. The set of seven plates, of which six are graduated on both sides, and the last on one side only, gives a complete scheme of old Japanese time. Needless to say that it agrees in no sense with our European system, but the subject can be dealt with more conveniently when I come to describe a final phase of pillar clocks, which dispenses with "hour" plates and their periodical adjustment.

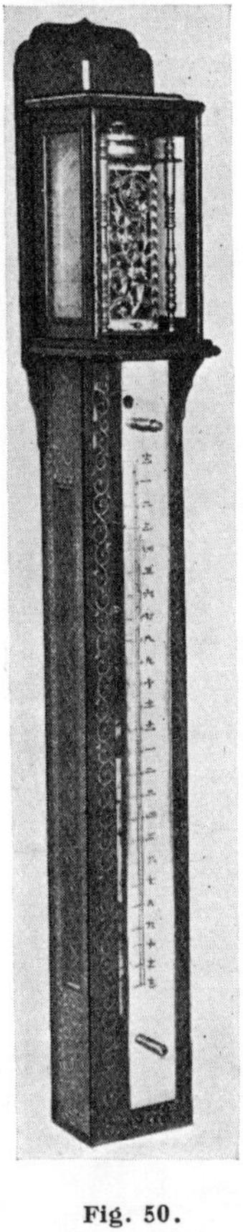

Fig. 50.

In the clock which we have been considering, all that was necessary was to fit the scale corresponding with the latter half of our December, and a fortnight later to reverse it for the first half of our January, and next to insert the succeeding scale, and so on, until the last scale representing the second half of June was reached, graduated on one side only. Then successively the other scales were used again in reverse order. It is as well to remark here that almost all the clocks of this description that have survived are fitted with a single scale only, the other members of the set having been discarded by the curio dealers as of no interest. My own collection contains only one complete set in its original box.

With the introduction of European time, it occurred to the Japanese clockmaker that these clocks might be adapted to the new system, and accordingly we find some specimens in which, instead of the Japanese set of scales, a single scale has been substituted, side by side with the old "hour" plates, which shows the time in equal hours commencing with twelve noon, and graduated in divisions of one-fifth through all the twice-twelve hours of our clocks. This is well shown in Fig. 49 (height $15\frac{3}{4}$ inches).

Fig. 50 shows a simpler way of dealing with European hours, by the substitution of a brass scale starting with 12 (±) for the old "hour" slab; or it may be that it was so made originally. It is a beautifully finished striking-clock, the case being ornamented with scroll-work in lacquer, and adorned with the official Imperial badge, the chrysanthemum (height 25¼ inches).

We now come to a very interesting and ingenious development in the pillar clock—the addition of striking works. How was this to be accomplished? for in the very narrow space allotted to the going train there was no room for any extra mechanism. If such an addition had been attempted, the whole build and symmetry of the clock must have undergone radical change. By some happy inspiration the possibility was conceived of utilizing the striking train as the motive power of the going train, in place of the leaden weight. The striking train was therefore enclosed in a brass case, made of such breadth and width as to permit of its descent within the wooden body of the clock. Generally speaking the weight by itself was insufficient, and it was augmented by screwing thick flat pieces of lead to the back or bottom of the brass casing.

The striking action was effected in this way. The lifting piece and locking plate, similar to those employed in the generality of bracket clocks, were placed outside and in front of the case containing the wheelwork with the bell at the bottom. I have described how the "hour" plates were fixed in the scale of the non-striking clock. The same method was used in the striking-clock with the difference that, as in the case of the bracket clock, the projecting piece was longer, and of such a length that when the weight (the striking train) in its descent came into line with an "hour" plate the projection raised the lifting-piece, which thereupon, with the locking plate correctly set, released the train, which then gave on the bell the number of blows corresponding with the particular numeral on the hour plate. In like manner one blow was struck at the half-"hours" following an uneven "hour" and two blows after an even "hour."

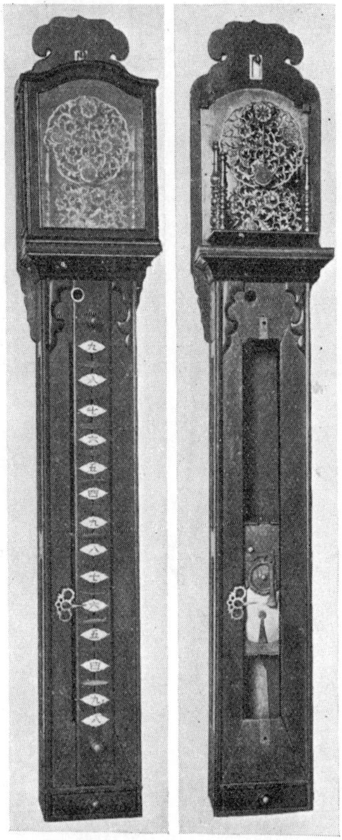

Fig. 51.

In order to wind the striking train it was necessary first to wind the going train by the square on the great wheel to its full extent, whereupon the square of the striking mechanism came opposite to a hole in the wooden case, through which its winding key was inserted.

Whether the new day was deemed to commence at dusk or, as in later clocks, at noon, it is obvious that some margin of time had to be allowed; for it could not be expected that the owner should be present at the precise moment of the change.

This difficulty was overcome by the addition of an extra hour plate (or in some clocks with the half-hour plate as well) at the bottom of the scale.

In the case of the non-striking clock the adjustment was easy. All that was necessary was to note the position of the pointer below the bottom plate, and then to wind up the weight so that the pointer should mark the same position below the top plate. But with a striking-clock the matter was not quite so simple. Let us suppose the ordinary case of a clock commencing the new day with the 6-plate, and having an additional 6-plate at the bottom of the scale. I have explained that the weight had to be wound up to its full extent to make the winding of the striking train possible. If then the clock had struck six at the bottom plate, it is obvious that it would strike again when the pointer was opposite to the top plate, and the number of the strokes would be the next in order, namely five, thus upsetting the whole sequence of the "hours." In order to prevent this happening, the top "hour" plate and the following half-"hour" plates (where the clock was furnished with half-"hour" divisions) were dummies. That is to say, they were pinned to the scale as in the case of non-striking clocks without any projecting piece at the back, and were therefore unable to raise the lifting-piece. As, however, the weight had to be wound to the top for windings, the further difficulty presented itself that the pointer would almost always have descended more or less below the lowest extra plate or plates. To obviate this the winding mechanism of the going train was in some cases so arranged that, after it had been wound by turning the key to the right, the action could be reversed, and turning to the left the pointer could be brought to the same posi-

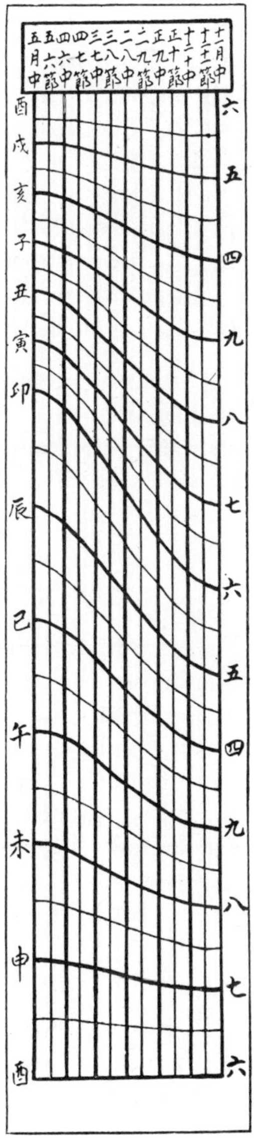

Fig. 52.

tion below the top plate that it had already reached below the bottom plate.

The illustration shows a very beautiful specimen of this type of clock, the front plate of which has been given in Fig. 42 (p. 265). On the left the clock is shown in its complete form. The " hour " scale commences with 9 (noon), and it will be seen that 9 and 8 are duplicated at the bottom. The figure on the right reproduces the clock with the hood and the scale removed, so as to show the striking mechanism, enclosed in a brass case, acting as the driving weight (Fig. 51).

The last stage in the development of the pillar clock was an ingenious device which obviated even the necessity for changing the interchangeable scales bi-monthly. This consisted in combining in a single scale all the varying " hour " divisions as shown on the thirteen sides of the complete set of seven scales, by means of a plotted scale, which is best explained by the accompanying diagram (Fig. 52).

At the top of the scale appear in the two top rows the numbers of the respective months, and in the bottom row signs indicating the commencement of a month and the relative half-month.

Translating these signs into arabic figures and using B for the beginning of the month and ½ for the latter half, we get the following arrangement :

5	5	4	4	3	3	2	2	1	1	12	12	11
–	6	6	7	7	8	8	9	9	10	10	11	–
½	B	½	B	½	B	½	B	½	B	½	B	½

It must be borne in mind that, as the Japanese year commenced approximately a month later than the European year, as explained in the chapter on the Calendar, it is necessary to add 1 to each of the above numbers, in order to understand how the scheme worked.

Below the above figures are drawn seven vertical lines, commencing in each case below the ½ sign, with dots (or sometimes additional lines) between to mark the commencement of a month. Down each side of the outside vertical

a b c d

Fig. 53.

lines are placed the hour numerals, or in some scales on one side the signs of the zodiac corresponding with the twelve "hours." Across the vertical lines run plotted curves which, at the point where they intersect the former, show the length of the respective "night" and "day" hours for the particular half of the month from which the vertical line starts.

The time was read in this way. Outside the scale a small flat brass bar was fixed by tangs at each end into the leaden driving weight. This bar carries a small pointer (often missing), so fitted as to slide along it. All then that was necessary was to set the pointer to the line corresponding with the current half-month. As then the weight descended the pointer marked the "night" and "day" hours of their proper length in accordance with the month for which it was set. Fig. 53 illustrates these clocks.

Obviously, this type of clock did not admit of a striking train, there being no projecting pins to discharge a striking mechanism; but the simplicity of its action no doubt recommended it to those who could not afford a more elaborate and expensive timekeeper, and one which required no professional adjustment.

For a very long time this scale caused me and everyone—Japanese and others—to whom I submitted the problem very great perplexity. The late M. Planchon, in his book "L'Horloge" (*op. cit.*), professed to give what on the face of it seemed to be a simple explanation; but when I pointed out to him that the day and night "hours" were all shown of equal length in the second half of December (i.e. round about the winter solstice) instead of at the equinoxes, he could only exclaim, "Ah, je n'y pensais pas!"

At last I discovered the true explanation in the passage from Siebold's "Nippon,"[1] which in translation reads as follows:

> When reckoning according to *civil* hours the Japanese include both twilights in the natural *day*, and generally place the commencement of the morning dawn and the end of the evening twilight at that moment of time when one begins or ceases to be able to recognize a printed character held in the hand, or when in the morning the stars disappear, and when in the evening they appear once more.

[1] Vol. iv, part iii, p. 117, etc., in Appendix II.

Astronomically, the period of twilight is reckoned from the moment when the sun is 18° below the horizon (i.e. 72 minutes); but this in the latitude of Kyōto, Tokyo, etc. (approximately 35° N.), would be excessive. For the sake of giving a simple explanation of the scale I will assume a period of one hour each for twilight and dawn.

First let us convert the Japanese table at the top of the scale into its approximate European equivalents. The result is as under:

Ju	Ju	My	My	Ap	Ap	Mc	Mc	Fe	Fe	Ja	Ja	De
–	Jy	Jy	Au	Au	Se	Se	Oc	Oc	No	No	De	–
½	B	½	B	½	B	½	B	½	B	½	B	½

Now, if we examine the scale (Fig. 52), we see that the "night hours," beginning with the numeral 6, are very short and the "day hours" very long in the second half of June, corresponding with the summer solstice, as they would be; whereas the hours on the right-hand side of the scale are all of equal length in the latter part of December.

Let us now suppose that in about latitude 35° N. the sun rises in midsummer at 5 and sets at 7 European time, and in midwinter rises at 7 and sets at 5. That is to say, that at the summer solstice the period between sunrise and sunset is fourteen of our hours, and at the winter solstice ten hours. But, as we have to include one hour each for dawn and twilight in the length of *daylight*, it results that we get sixteen hours of light and eight hours of darkness in midsummer, and *twelve* hours for *each* in midwinter, each period of light and darkness being divided into six Japanese "hours" of equal length. This then was the principle on which the plotted curves were drawn.

Assuming that the clock were started half-way through December the pointer would be set against the vertical line on the extreme right.[1] At the beginning of January it

[1] Occasionally the scale is drawn in the reverse sense with the equal hours commencing on the left side.

would be moved towards the next line, and again in the middle of January to the third line from the right, and so on until the line representing the latter part of June on the extreme left was reached, all these readings corresponding with the top row of month numerals. The process would then be reversed, reading from left to right along the second row until mid-December was reached once more. As a consequence of this method of always adding an hour for dawn and twilight to the daylight period, darkness at the equinoxes lasted approximately ten European hours and light fourteen hours, instead of the two periods being, according to our reckoning, of equal duration.[1]

As a general rule the scale of these clocks is of wood, lacquered in black, with the figures in gold, as shown in Fig. 53, but one specimen in my collection has a scale of brass with the figures engraved. Its general appearance suggests an early effort, the front plate being neither engraved nor pierced[2] (Fig. 53 b).

And here it is worthy of more than passing mention that Prof. A. A. Rambaut,[3] in a paper, read on February 20, 1889, before the Royal Dublin Society (recorded in vol. vi, p. 332, of their "Scientific Proceedings"), entitled a "Note on Some Japanese Clocks," succeeded by pure astronomical reasoning in solving this knotty problem, unaided by any Japanese information. The chance discovery of this paper in the Library of the Victoria and Albert Museum enables me to pay this tribute to Prof. Rambaut's clever induction, which so completely confirms Siebold's statement.

[1] The above explanation applies to the latitude of Kyoto (35° N.). I possess a pillar clock of the ordinary type on the brass scale of which is engraved the variation of the length of the divisions for the latitude of Nagasaki (33°).

[2] I have recently seen a brass scale which served to actuate a striking train. The lines on the front are oblique and not curved, corresponding lines being deeply cut on the back, so that the projection behind the hour-plate on meeting with this groove set off the striking detent. Through the absence of the curves the time would only be given approximately.

[3] Astronomer Royal of Ireland, and (1927) Astronomer of the Radcliffe Observatory.

Fig. 54.

Fig. 55.

Fig. 56.

Fancy Clocks

Apart from the three main categories of clocks so far described, the Japanese craftsman indulged his fancy in the production of a variety of pieces, of which many were made in imitation of European models. Such clocks, as they follow no general rule, can best be described by illustration of such as find a place in my collection.

Fig. 57.

1. *Table Clocks*

(Figs. 54 and 55)

The clocks from which these were copied were mostly of German origin, made in the seventeenth century. As the name implies, they stood on a table or similar support, necessitating a horizontal dial. Fig. 54 measures $2\frac{3}{8}$ inches and Fig. 55, 2 inches without the stand.

2. *Silver Clock-watch*

(Fig. 56, diameter $3\frac{1}{4}$ inches)

These portable clocks, in form like a large watch, were also more frequently made abroad than in England. The back is beautifully engraved with a basket of flowers. The specimen illustrated has the travelling case of Shitân wood.

3. *Paper-weight Clock*

The little gem here illustrated (Fig. 57) has been described in a Japanese paper as a " doctor's clock." However this may be, it would certainly answer the purpose of a watch.

The outer case of Shitân wood[1] measures no more than $5\frac{3}{8} \times 1\frac{1}{8}$ inches, whilst the clock itself measures only $4\frac{3}{4} \times \frac{13}{16}$ inches. The mechanism reproduces the design of a pillar clock ; but, inasmuch as a portable clock could not be driven by a weight, a spring action is employed. The spring is contained in a barrel, housed in a compartment at the bottom of the case. As the clock is wound, bringing the pointer to the top of the " hour " scale, the spring is tightened. Then, as it gradually uncoils, the pointer is drawn downwards and marks the " hours."

The circular balance with hair-spring at the top is fitted with a regulating needle, similar to that of a European watch. The case is of gilt brass, and the whole piece is enclosed in an outer casing of decorative fabric.

Fig. 58.

4. *Inro*

It is not surprising that the *inro*, or medicine box, should have suggested a means of providing a portable timepiece. A case of this description containing clockwork, and attached to the girdle by its *netsuke* and *ojime*, would serve the purpose of a pocket watch, for which the kimono provided no receptacle.

Of the three specimens illustrated (Fig. 58) one is a little *striking* clock enclosed in a Shitân wood case adorned in colour and gold with maple leaves, chrysanthemum

[1] Lid removed for photographing.

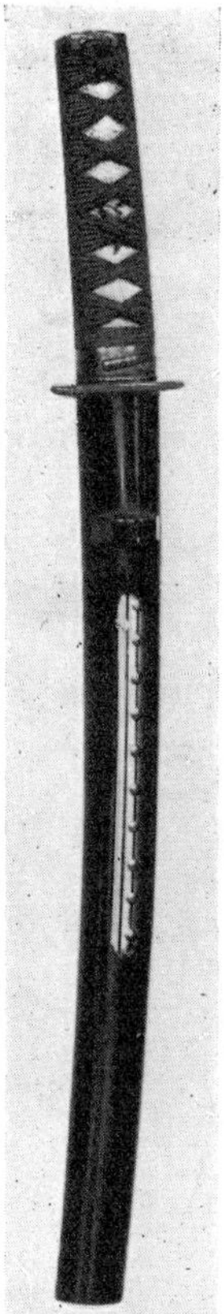

Fig. 59.

flowers and butterflies, the *ojime* being decorated with a spray of leaves and coral fruit, and having a mother-of-pearl cartouche with the signature *Furuya.*

The second, from the Michael Tomlinson collection of Japanese works of art (26.6.22), is thus described in the sale catalogue: "Shibuichi case (i.e. blackened silver) engraved with flowers, containing a Japanese clock, key contained in manju ; case signed *Asahi.*" In the bottom of the case a compartment contains a compass.

The third example is a little clock enclosed in an old stamped leather case, with a flap in front fastened by a prototype of the modern press-button. The little silver *ojime* is signed *Tani.* The ivory *netsuke* represents a kneeling man with a baby on his back. A compass is let into the back plate of the movement.

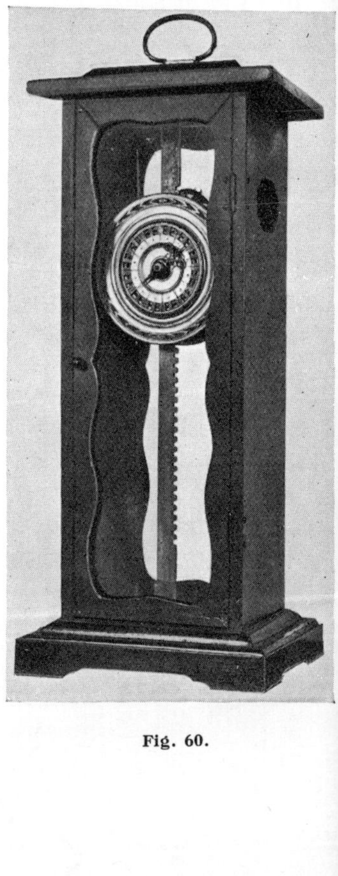

Fig. 60.

5. *Sword Clock*

This very interesting piece is another example of a clock to be worn on the person.

The short sword was permitted to be carried by certain persons who were not of *noble* rank, amongst whom might be included a doctor, attached as a retainer to his feudal lord.

The clock illustrated in Fig. 59 has a scabbard of brown lacquer, which through a slot on one side shows the scale of the clock, the mechanism being placed in the handle. The action is similar to that described under No. 3 (p. 281). If in fact it was owned by a doctor, it is possible that it was for use on occasions when he attended his chieftain on some military expedition (length 21½ in.).

Fig. 61.

6. *Rack Clock*

This clock (Fig. 60) is another imitation of a European design. The circular case containing the going and striking trains acts by its own weight as the motive force on its descent down the vertical rack (height 14 inches).

7. *Octangular Striking-Clock*

The chief interest in this clock (Fig. 61) lies in its unusual form. It may be described as a combination of the latest type of lantern clock, with circular balance and revolving "hour" circle, and the pillar clock, with the "hours" marked by a pointer attached to the descending weight, which drives both trains. The plates in the latter case are engraved with the signs of the zodiac. Thus, by this arrangement, both methods of reading the time are provided for (height $25\frac{1}{4}$ inches).

CONCLUSION

THE preceding chapter has been illustrated entirely from clocks in my own collection ; but it is needless to say that, whilst the standard types followed the same general design, the detail of the cases and the external ornamentation, and to a lesser extent the mechanism of domestic clocks were the subject of an infinite variety of treatment at the hands of the artistic craftsman.

There are not many specimens in the London Museums ; but a few will be found in the Victoria and Albert Museum and in the Science Museum, some in each case being on loan by Mr. P. L. Harrison, a collector now resident in Yokohama, to whom I am indebted for a translation of Mr. Takabayashi's book (*op. cit.*).

Mr. Takabayashi states that clocks are mentioned in works of poetry at the end of the seventeenth century. He adds that designs of various clocks were exhibited in shops, and that they were made in Nagasaki, Kyōto, Osaka, Yedo, Sendai, etc. ; but how far back the work of the clocksmith, as he was called, became a recognized craft does not appear ; and the absence of signature and domicile leaves the matter in doubt.

The makers of whom he speaks with certainty belong to the end of the eighteenth century and thereafter. We have seen that Hosokawa wrote his treatise on the construction of a " lantern " clock in 1796. It appears that one Asada about the same time constructed a form of balance clock controlled by a long pendulum and light bob swinging at the side of the case, which had two crown-wheel escapements. The illustration shows that it had two dials. The larger one has an outer circle, numbered 1 to 10, each division being again divided into tenths, and an inner disk similarly divided. It has two hands, the smaller one revolving round the disk and the longer following the outer circle. The smaller dial below with its single hand follows the same divisions ; so that presumably it was a decimal counter, dividing each numeral on the disk into tenths, hundredths, and thousandths. It was used by the astronomer Tadanori Ino, and is a treasured possession of his descendants.

Another astronomical clock is figured in diagram in Mr. Arthur James's article in *Popular Mechanics* (Chicago, September 1909), which, beside lunation work, gives the varying " hours " on the Japanese system. Such clocks are of extreme rarity.

In the 1830's we are told by Takabayashi that there were clocksmiths in different cities who produced clocks of novel shapes ; and he speaks of the beautiful finish of these hand-made clocks as an indication of the art and skill which the Japanese craftsman had attained. This statement confirms my theory that it was at this period that the small " lantern," the " bracket " and the " pillar " clocks were evolved.

During this period it was that a famous maker of mechanical devices, Denjiro Kobayashi, of Tokyo, made a musical clock, which externally bears a resemblance to my Fig. 32.

The daimyōs at that time, says Takabayashi, " bought whatever new designs were produced, and the clockmakers did their utmost to construct as many artistic and carefully thought-out designs as possible." Matsudaira, the Prefect of Dewa, it seems, had a different form of clock for each room, which he designated by the name of the clock that it contained.

One of the greatest of the clockmakers of last century was Hisashige Tanaka (1798–1881). In 1850 he made a 400-day clock with six dials, showing respectively : (1) European time, (2) Japanese time, (3) the days of the week, (4) the twenty-four seasons, (5) lunation, (6) the sexagenary cycle. In addition it had on top a mechanism showing the sun's rotation. It was exhibited at the Horological Exhibition in Tokyo in 1920, and is still preserved by the Tanaka family.

The names applied by the Japanese to their various types of clock were not such as, for convenience, I have adopted. I therefore append the denominations as given by Mr. Takabayashi :

1. Wall clock	*i.e.*	Suspended " lantern " clock.
2. Pyramid clock	.	" Lantern " clock on conical stand.
3. Table clock .	.	Small " lantern " clock on stand.
4. Pillow clock	.	" Bracket " clock (so-called from its resemblance to a Japanese pillow).
5. Foot-rule clock	.	" Pillar " clock.
6. Flat-top clock	.	Table clock.
7. Inro clock.		
8. Paper-weight clock		Doctor's clock or watch.
9. Watch clock	.	Clock-watch.
10. Taper clock	.	Clock with luminous dial.
11. Ship's clock.		
12. Bell clock .	.	Mechanism in a case like a temple bell.

With the new era of European time the first factories were installed, the first in 1875. The most successful—the Seikosha factory, established in 1888—was totally destroyed by the great earthquake of 1923, involving a loss of over £1,500,000.

BIBLIOGRAPHY

In order to facilitate research I have indicated by an initial the public libraries in London, when known to me, where a work may be consulted, thus :

B = British Museum ; G = Guildhall ; P = Patent Office ; S = Science Museum ; V = Victoria and Albert Museum.

There is also the Vulliamy Collection of books and pamphlets at the Institute of Civil Engineers, which contains nothing after 1854, but up to that date is voluminous. Other works have been acquired since. These are marked (I.).

In regard to the British Museum it may be assumed that all works published since the coming into force of the Copyright Act 1842 (July 1 of that year) are to be found there ; and for such work the initial letter was unnecessary.

In the case of works having titles of inordinate length, I have given only so much as is necessary to identify the book and its scope, in order to economize space. I have added, as far as I have been able, particulars of the format, number of pages, illustrations, place of publication and its date.

Abbott, Francis

A Treatise on the Management of Public Clocks. 8°. 44 pp., 4 pl. (G.) *London*, [1835].

Second edition. (P.)

Third edition. 46 pp. (P.) [1839.]

Abbott, H. G.

Watchmaker's and Jeweler's Practical Handbook. 118 pp. Five editions. *Chicago*.

The Watch Factories of America Past and Present. 8°. *Chicago*, 1888.

The American Watchmaker and Jeweler (Encycl.). 354 pp., 288 illus. (Also later editions.) *Chicago*, 1892.

Antique Watches and how to establish their Age. 8°. 204 pp., numerous illus. (B.) *Chicago*, 1897.

The Roberts Collection of Antique Watches. *Chicago*, 1897.

Académie des Sciences

Report approving the New Escapement for Watches of Caron fils. 12°. 40 pp. (P.) *Paris*, 1754.

Report on One-wheeled Clock of le Mazurier. 12°. 24 pp. (P.) *Paris*, 1755.

Airy, Sir G. B.

On a New Construction of Fusee. (I.) 1840.

Description of Airy's Clock Escapement. (I.) 1842.

Albèri, E.

Intorno ad una disquisizione storica circa la prima applicazione del pendolo all' orologio. 4°. [1858.]

(A reply to J.-B. Biot's discourse on the Galileo clock.)

Albert, J. F.

Manuel d'horlogerie. 8°. 1909.

Alléaume, A.
Les brevets d'invention (français) concernant l'horlogerie, etc. 8°. 144 pp. (P.) *Paris*, 1873.
Allexandre, Jacques
Traité général des horloges. 8°. 392 pp., many copper plates. (B., G., P.) *Paris*, 1734.
(A most interesting early work on practical horology; it contains a descriptive bibliography.)
For *German translation* see Berger.
Almanacs
Annuaire artistique et historique des horlogers. Edited by Cl. Saunier. Commencing *Paris*, 1859.
Almanach des horlogers. Commencing 1886.
Notizkalender für Uhrmacher. Commencing 1878.
Ambronn
Handbuch der astronomischen Intramentenkunde. 4°. *Berlin*, 1899.
Andrade, J.
Chronométrie. 12°. 1908.
Les organes réglants de chronomètres. 8°. 150 pp. (P.) *Besançon*, 1922.
Horlogerie et chronométrie. 8°. 582 pp. (P.) *Paris*, 1924.
Anonymous Works
Excellency of Horology. (I.) 1689.
On the Longitude . . . in a Letter to the Commissioners of that Board. (I.) 1781.
L'Art de l'horlogerie enseigné en trente leçons . . . d'après Berthoud et les travaux de Wulliamy . . . par un ancien élève de Breguet. 12°. (G.) *Paris*, 1827.
Second edition. 2 vols. *Brussels*, 1828; Third edition. 12°. 1835.
For *German translation* see Wolbrecht.
Mémoires sur l'horlogerie, contenant diverses remarques sur les ouvrages et les prétentions de M. R. (? Robert Robin). 4°. 40 pp., 2 pl. (P.) *Paris*, 1793.
Anweisung für Uhrmacher und sonst Liebhaber genauer Uhren, wie selbe nach dem Sonnenlaufe können gerichtet werden, etc. 12°. *Constance*, 1826.
Description des échappements les plus usités en horlogerie, etc. *Geneva*, 1831.
Anquetin, Georges
Eléments usuels et pratiques du réglage des montres.
Anquetin, Modeste
L'Horlogerie : des montres en général : de ceux qui les font, etc. 12°. (B.) *Paris*, (?) 1875.
Eléments de repassage en second pour conserver les montres, etc. 16°. 43 pp.
Antoine, E.
Notes sur le réglage. 8°. 32 pp. *Paris*, 1891.
De la méthode en réglage. 8°. 152 pp., 2 pl. (P.) *Geneva*, 1902.

ARAGO, see Gay-Lussac.

ARENTZ, F. C.
Beschreibung der astronomischen Uhr . . . von Herrn N. A. Johann 1807 berechnet und verfertigt, etc. 4°. 4 pl. *Mayence*, 1829.
Second edition. 1830.

ARNOLD, John
An Account kept during 13 Months at Greenwich Observatory of the going of a Pocket Chronometer, etc. 4°. 12 pp. *London*, 1780.
Tracts and Articles (see Catalogue of the Clockmakers' Company). (G.)
(See also C. Mayer.)

ARNOLD, J. R.
Explanation of Timekeepers constructed by Mr. J. Arnold. 4°. 22 pp., 3 pl. (P.) *London*, 1805.
(See also Earnshaw and Arnold.)

ARTHUR, James
Time and its Measurement. (Articles in August to November numbers of *Popular Mechanics*.) *Chicago*, 1909.

ARZBERGER, C.
Anleitung zur Berichtigung des Ganges der Uhren. 8°. *Coburg*, 1802.

ARZBERGER, F.
Die elektrische Uhr. 8°.

ATKINS and OVERALL
Some Account of the Clockmakers' Company. 8°. (G., P.) *London*, 1881.

ATWOOD, G.
Investigations founded on the Theory of Motion for determining the Time of Vibration of Watch Balances. 4°. 52 pp., 2 pl. (P.) *London*, 1794. (In the *Philosophical Transactions*.)

AUCH, J.
Taschenbuch für Uhrenbesitzer, etc. 8°. 460 pp., 8 pl. *Gotha*, 1806.
Second edition. 1808.
Anleitung zur Kenntniss u. Behandlung der Taschenuhren für Uhrenbesitzer, etc. *Gotha*, 1808.
Handbuch für Land-Uhrmacher, etc. 8°. 3 pl. *Ilmenau*, 1827.
Second edition. 12°. 261 pp. *Weimar*, 1858.
Third edition. 16°. 104 pp. (P.) *Weimar*, 1892.

AUGUSTIN (Père Religieux)
Horologéographie pratique, ou la manière de faire les horloges à poids. 8°. (Described by Allexandre, p. 350.) *Rouen*, 1719.

AUHL, A. M.
Gemeinfassliche Anleitung zur Beurtheilung, Behandlung, Regulierung und Reparatur aller Arten von gewöhnlichen Uhren, etc. 16°. 100 pp. *Vienna*, 1851.

AXON, W. E. A.
Receipts and Practical Suggestions relating to Horology. (Article in *The Mechanic's Friend*, pp. 150–75.) 8°. (P.) *London*, 1875.

BABEL, A.
Histoire corporative de l'horlogerie. 606 pp. (B.) *Geneva*, 1916.

BACHELIN, A.
L'Horlogerie Neuchâteloise. 8°. 224 pp., 1 pl. (B.) *Neuchâtel*, 1888.

BAILLIE, G. H.

Watches: their History, Decoration and Mechanism. With numerous illustrations in Colour and Collotype. 8°. 408 pp., 76 pl. (P.) *London*, 1929.

Watchmakers and Clockmakers of the World. 8°. *London*, 1929.

Public Clocks erected before 1600. (In the *Watch and Clockmaker*.) *May* 1931.

BAIN, A.

A Short History of Electric Clocks, etc. 8°. 32 pp. (P.) *London*, 1852.

(See also Barwise.)

BAJOT, Ed.

Encyclopédie du meuble du XVe siècle, etc. (Contains particulars of clocks.) (V.) *Paris*, 1901– .

BALAVOINE, F.

L'échappement à détente. 8°. 1904. Second edition. 8°. 48 pp. (P.) *Geneva*, 1912.

BALL, A. E. J., see Langman.

BANKS, Sir J.

Protest against granting a Reward for Earnshaw's Chronometers. (I.) 1804.

BARFUSS, Fr. W.

Geschichte der Uhrmacherkunst von den ältesten Zeiten bis auf unsere Tage. (B.) 1837.

(Re-edited by Em. Schreiber and E. Gelcich, *q.v.*)

BARGONI, C.

Difetti dell' Orologio, etc. 1797.

BARKER, A.

How to examine and clean Clocks. 8°. 8 pp. *Stockton*, 1903.

BARNARD, F. A. P.

On the Pendulum. 8°. 20 pp. (P.) *New York*, 1859.

BARRAUD.

Cloches des horloges. (In *Annales Archéologiques*, vol. xviii.) 1858.

BARRAUD, H.

Patent Correcting Weight for Chronometers, etc. 8°. 16 pp., 1 pl. (P.) *London*, 1848.

Chronometers. (I.) *London*, 1850.

BARRÈRE, G.

Traité élémentaire d'horlogerie à l'usage des horlogers rhabilleurs. 8°. 1892.

BARRINGTON, S.

Remarkable Musical and Other Clocks, including the Westminster Clock of 1288. (In *Archæologia*, vol. v, p. 416.) *London*, 1779.

Letter to Blackstone as to Clock. (In *Annual Register*, p. 133.) *London*, 1779.

BARWISE, J., and BAIN

Electro-magnetic Clocks. (I.) *London*, 1841.

BASSERMAN-JORDAN, Ernest von

Die Geschichte der Räderuhren unter besonderer Berücksichtigung der Uhren des bayerischen National Museums. Fol. edition de luxe (B.) 4°. 113 pp., many illustrations. (P., V.) *Frankfort a/M.*, 1906.

Katalog einer Sammlung von Goldemailuhren, vorwiegend des XVIII Jahrhunderts, etc. 4°. (B.) *Munich*, 1912.

BASSERMAN-JORDAN, Ernest von—*cont.*

Uhren. Ein Handbuch für Sammler und Liebhaber. 8°. 162 pp., 110 illus. (B., V.) *Berlin*, 1914.

(Vol. 7 of Bibliothek für Kunst und Antiquitätensammler.)

Second edition. 174 pp. *Berlin*, 1920.

Katalog einer Uhrensammlung aus süddeutschem Privatbesitz, vorwiegend Taschenuhren und Halsuhren des XVI–IX Jahrhunderts. 4°. 52 pp. *Munich*, 1917.

Die Geschichte der Zeitmessung und der Uhren. 4°. (P., S.) *Berlin*, 1920– .

Alte Uhren und ihre Meister. 4°. 188 pp., 4 pl. (P.) *Leipsic*, [1926].

BASTIUS, Faustino.

Manuel del Relogero. 8°. *Paris*, 1864.

(In the *Enciclopedia Hispano-Americana.*)

BAUDON, A. and E.

Nouvelle notice sur la collection de cadrans de montres appartenant à M. C. Roblot. 8°. (B.) *Paris*, 1904.

BAUER, B.

Abhandlung von Taschenuhren. 8°. *Buchholz*, 1786.

BAUER, J. B.

Hemmungen und Pendel für Präcisionuhren, etc. 1894.

BAUER, J. C.

Abhandlung von der richtigen Beschaffenheit . . . und gehörigen Behandlung der Taschenuhren. 46 pp. 1798.

BECHER, J. J.

De nova temporis dimitiendi ratione et accurata Horologiorum constructione theoria et experientia. 4°. (B.) *London*, 1680.

BECKETT-DENISON, Edm. (Lord Grimthorpe)

Turret Clock Remontoirs. (I.) 1849.

A Rudimentary Treatise on Clock and Watchmaking. 8°. 280 pp. (G., P., S.) *London*, 1850.

Fifth edition. (P.) 1868.

A Rudimentary Treatise on Clocks and Watches and Bells. (Sixth edition of above.) (G., P., S.) *London*, 1874.

Seventh edition, revised and enlarged. 400 pp. (G., S.) *London*, 1883.

Eighth edition. 404 pp. (P.) *London*, 1903.

Also articles on Clocks in the eighth and ninth editions of the *Encyclopædia Britannica*, the latter (1857) containing a full account of the Great Clock at Westminster ("Big Ben"). For *German translation*, see C. H. Schmidt.

BECKMAN, Ferd.

Beyträge zur Geschichte der Erfindungen. Second edition. 1785.

English Translation in "History of Inventions," vol. iii, containing Hamberger's Treatise on Horology. *London*, 1797.

(See also Griffith.)

BEILLARD, Alfred

Petit recueil historique de l'horlogerie, ses progrès, etc. 12°. (B.) *Pacy-sur-Eure*, 1882.

Notes sur l'horlogerie française à l'Exposition Universelle de 1889.

Recherches sur l'horlogerie, ses inventions et ses célébrités, etc. 8°. 208 pp. (B., G., P.) *Paris*, 1895.

BELGRANO, L. T.
Degli antichi orologi d'Italia. (In *Arch. Storico Italiano.*) *Florence*, 1868.

BÉLIARD
Réflexions sur l'horlogerie en général et sur les horlogers du roi en particulier. 1767.

BELLEMO, Vicenzo
Jacopo e Giovanni de' Dondi dall' Orologio. 1894.

BENNET, J.
On a New Metallic Alloy for the Pivot Holes of Watches. 8°. 16 pp. (P.) *London*, 1830.

BENSON, J. W.
Illustrated Pamphlet of Clocks. 8°. 24 pp. (P.) *London*, 1862.
The Illustrated Pamphlet of Watches and Clocks. Tenth edition. 8°. 64 pp. (P.) *London*, 1864.
Time and Timetellers. 8°. 198 pp. (G., P.) *London*, 1875.
Second edition. 8°. 144 pp. (B.) *London*, 1902.
Turret Clocks. 8°. 70 pp. (P.) *London*, 1882.

BERGER, C. P.
Ausführliche Abhandlung von den Uhren. (*German translation* of Allexandre, *q.v.*) 8°. 400 pp. *Lemgo*, 1738.
Second edition. (P.) *Lemgo*, 1763.

BERLIOZ
L'Horlogerie dans toutes ses parties, etc. 8°. 1876.
Rapport sur l'horlogerie à l'Exposition de 1878.

BERNOUILLY, Daniel.
Recherches mécaniques et astronomiques sur la meilleure manière de trouver l'heure en mer. (In *Mémories de l'Académie des Sciences.*) *Paris*, 1747.

BERTHOUD, Ferdinand (1727–1807)
L'Art de conduire et de régler les pendules et les montres. 16°. 96 pp., 4 pl. (P.) *Paris*, 1759.
Second edition. 12°. (G.) *The Hague*, 1761.
Third edition. *Paris*, 1805.
Fourth edition. *Paris*, 1811.
Fifth edition. 16°. 5 pl. (P.) *Paris*, 1828.
Sixth edition. 12°. *Brussels*, 1836.
Essai sur l'horlogerie, etc. 2 vols. 4°. 490 + 452 pp., 35 pl. (P.) *Paris*, 1763.
Second edition. (G., P., S.) *Paris*, 1786.
Traité des horloges marines. 4°. 630 pp., 27 pl. (G., P.) *Paris*, 1773.
Eclaircissemens sur l'invention, la théorie, la construction . . . des nouvelles machines . . . pour la détermination des longitudes en mer par la mesure du temps. 4°. 162 pp. (G., P.) *Paris*, 1773.
Les longitudes par la mesure du temps, etc. 4°. (G.) *Paris*, 1775.
Mesures du temps appliquées à la navigation, etc. *Paris*, 1782.
De la mesure du temps, ou supplément au Traité des horloges marines et à l'Essai sur l'horlogerie. 4°. 275 pp., 11 pl. (G., P.). *Paris*, 1787.
Traité des montres à longitude, etc. 4°. (G.) *Paris*, 1792.
Suite du Traité des montres à longitude. *Paris*, 1796 & 1797.
Supplément au Traité des montres à longitude. 88 pp. (G.) *Paris*, 1807.

BERTHOUD, Ferdinand—*cont.*

De la mesure du temps par les horloges. *Paris*, 1797.

Histoire de la mesure du temps par les horloges. 2 vols. 4°. 374 and 447 pp. (B., G., P., S.) *Paris*, 1802.

Appendix to last work, vol. 3, 23 pl., no text. *Paris*, 1802.

Translations of Berthoud's Works.

Anweisung die einfachen sowohl als Repetir-Uhren wohl zu richten und zu gebrauchen. 8°. 28 pp. ("L'Art de conduire.") (P.) *Dresden*, 1759.

Anweisung zur Kenntniss, guten Gebrauchs und Haltung der Taschenuhren. 8°. 118 pp. *Meissen*, 1791.

Second edition. *Meissen*, 1818.

Die Kunst mit Pendel- und Taschenuhren umzugehen und sie zu regulieren, etc. (Translation of the Fifth edition.) *Weimar*, 1828.

Die Kunst die Pendel- und Taschenuhren zu behandeln, etc. (Translation of the Sixth edition of 1836 by F. Menadier.) 12°. 5 pl. *Quedlinburg*, 1838.

Versuche, Vortheile, Grundsätze und Regeln zur Erreichung der möglichsten Vollkommenheit der Taschenuhren, etc. (Translation and comments, by Fr. Vogel.) 8°. 250 pp. *Meissen*, 1790.

Italian translation: L'Arte de adoperare e regolare gli Orioli a pendolo e da tasca. 18°. *Florence*, 1778.

BERTHOUD, F., and JANVIER, L.

Nouveau manuel complet du régulateur des horloges, des montres et des pendules. 12°. 232 pp. (P.) *Paris*, 1838.

BERTHOUD, Louis

Entretiens sur l'horlogerie à l'usage de la marine. 127 pp. *Paris*, 1812.

BESSEL, F. W. (Astronomer)

Untersuchungen über die Länge des einfachen Secundenpendels. 4°. 2 pl. *Berlin*, 1829.

New edition, by H. Bruns. 171 pp., 2 pl. (B.) (In Ostwald's Klassiker.) 1889.

Bestimmung der Länge des einfachen Secundenpendels für Berlin. *Berlin*, 1835.

Abhandlungen. 4°. (Horological articles in vols. 2, 3, etc.) 1876.

BIAIS, E.

Notes sur l'horlogerie à Angoulême. (In the *Bulletin de la Société Archéologique et Historique de la Charente.*) (V.) *Angoulême*, 1909.

BILFINGER, Gustav

Antike Stundenzählung. 4°. 41 pp. *Stuttgart*, 1883.

Die Zeitmesser der antiken Völker. 4°. 78 pp. *Stuttgart*, 1886.

Die babylonische Doppelstunde. 4°. 55 pp. *Stuttgart*, 1888.

Die antiken Stundenangaben. 8°. 169 pp. *Stuttgart*, 1888.

Der bürgerliche Tag, etc. 8°. 296 pp. *Stuttgart*, 1888.

Die mittelalterlichen Horen und die modernen Stunden. 8°. 289 pp. (B.) *Stuttgart*, 1892.

BION, N.

Traité de la construction et usage des instrumens de mathématique. (Book VI, chap. 5, reproduces Huygens' clock in "Horologium Oscillatorium.") 4°. (Second edition.) *Paris*, 1716.

New edition. *The Hague*, 1723; Revised edition. 4°. *Paris*, 1725.

BION, N.—*cont.*

English translation, with additions by E. Stone. The Construction and Principal Uses of Mathematical Instruments. Fol. (B.) *London*, 1723.

German translation, by J. G. Doppelmayr: Mathematike Werck-Schule. 4°. (B.) *Nuremberg*, 1726.

BLAVIGNAC, J. D.

La Cloche : étude sur son histoire. *Geneva*, 1877.

BOCK, H.

Die Uhr, etc. 8°. 136 pp. *Leipsic*, 1908.

Second edition. 121 pp. 1917.

Kritische Theorie der freien Riefler-Hemmung. 8°. 70 pp. (P., S.) *Berlin*, 1910

BÖHM, J. G.

Die Kunstuhren auf der k.k. Sternwarte zu Prag. 4°. 48 pp., 21 full-page illustrations. (B.) *Prague*, 1908.

BOHMEYER, A.

Der Proportionalzirkel für Gross- und Kleinuhrmacher. 20 pp., 2 illus. 1882.

BOHMEYER, C.

Anleitung zur Aufstellung elektrischer Uhren. 8°. 94 pp., 45 figs. (B.) *Hanau*, 1892.

Third edition. 8°. 138 pp. (P.) *Bautzen*, 1908.

BOISSY D'ANGLAS, F. A.

Rapport . . . sur la manufacture d'horlogerie à Besançon. 8°. 8 pp. (P.) *Paris*, 1794–5.

BOLTON, L.

Time Measurement. 8°. 174, pp., 8 pl. (P.) *London*, 1924.

BOND, W. C.

Observations in the Comparative Rates of Marine Chronometers. (I.) 1829.

BONNAFFÉ, E.

Preface to the Sale Catalogue of the Spitzer Collection of Clocks and Watches. *Paris*, 1893.

BOOTH, Mary Louise

New and Complete Clock and Watchmakers' Manual. (Taken from Le Normand and Janvier's work by Magnier, *q.v.*) 8°. 305 pp. (G., P.) *New York*, 1860.

Reprinted. 1882.

History of the Manufacture of Watches. 8°. 1877.

Clock and Watchmakers' Manual, etc. *Chicago*.

BORCHHARDT, L.

Die Geschichte der Zeitmessung und der Uhren.

BOREL, G.

Paris Exhibition, 1900. "Horlogerie": Rapport du jury international. (V.) *Paris*, 1902.

BOUHAIN

Description d'une horloge merveilleuse qui a été fabriquée dans la ville de Niort en Poitou par le Sieur Bouhain. 12°. (Quoted by Allexandre, p. 326.) (?)

BOURDAIS, M.

400 procédés modernes à l'usage des horlogers, etc. (?)

Second and third editions. 180 pp. *Paris*.

BOURDAIS, M., et GRALL, G.
A.B.C. de l'apprenti horloger. Second edition. 12°. 239 pp. *Paris*, 1917.

BOUVIN, F.
Les secrets de l'horlogerie. 16°. 48 pp., 4 figs. *Amiens*, 1891.

BOUYER, R.
La décoration des montres. (In *Art et Décoration.*) (V.) 1901.

BOWMAN, J. J.
The Manipulation of Steel in Watchwork.

BRASSLER, C. A.
A Practical Treatise on the Balance Spring. 12°. *New York*, 1876.

BREARLEY, H. C.
Time Telling through the Ages. 8°. 294 pp., 24 pl. (P.) *New York*, 1919.

BREGUET, Abraham Louis (1747–1823)
Important article in Laboulaye's *Dictionnaire des Arts et Manufactures.*

BREGUET, L. F. C.
Paris Exhibition, 1867. Rapport du jury international. Pp. 317–36. (P., V.)

BRIOT
Rapport sur la manufacture nationale d'horlogerie établie à Besançon. 8°. 6 pp. (P.) *Paris*, 1798.

BRITTEN, F. J.
The Watch and Clockmakers' Handbook, Dictionary and Guide. (The first edition was published in the early '70's and the 13th edition in 1922.) *London*, v.d.
On the Springing and Adjusting of Watches. 8°. 152 pp. (P.) *London*, 1898.
Some Old English Bracket and Long-case Clocks. (In *Magazine of Fine Arts*, p. 65 and p. 424.) (V.) 13 pp. *London*, 1905–6.
Old English Clocks : the Wetherfield Collection. Large 4°. 114 pp. *London*, 1907.
Former Clock and Watchmakers. 8°. 397 pp. (G., P., S.) *London*, 1894.
Old Clocks and Watches and their Makers. 8°. 500 pp. (S., V.) *London*, 1899.
Second edition, enlarged. (P., S., V.) 740 pp. *London*, 1904.
Third edition, much enlarged. 790 pp., 800 illus. (P., S., V.) *London*, 1911.
Fourth edition, Old Clocks and their Makers. 597 pp. (V.) *London*, 1919.
(Watches were omitted from this edition on account of high costs following the Great War.)
Fifth edition (with Watches and further enlarged). *London*, 1922.

BRITTEN, F. W.
Horological Hints and Helps. 8°. 340 pp. *London*, 1930.

BRÖCKING, W.
Ueber elektrische Uhren, etc. 12 pp., 1 pl. 1873.

BROCOT, A.
Calcul des rouages, etc. 8°. *Paris*, 1862.
German translation : Berechnung der Räderübersetzungen. 1871 and 1879.

BRÖNIMANN, F.
Ueber den Isochronismus des Pendels und der Unruhe. 1875.

BROOK, A. J. S.
Notes on a Brass Table Clock and a Silver-cased Table Clock. (In the *Proceedings of the Society of Antiquaries of Scotland.*) (V.) *Edinburgh*, 1906.

BROWN, A. D.
The Electrical Distribution of Time. (In the *Journal of the Franklin Institute.*) *Philadelphia*, 1888.

BROWN, Isaak
Die neuesten Erfindungen und Verbesserungen in der Uhrmacherkunst, etc. (A translation from an English work.) 8°. 352 pp., 138 figs., 36 pl. *Quedlinburg*, 1831.
Second edition, revised. *Quedlinburg*, 1854.

BROWNE, Robert
Method of finding the Longitude at Sea. *London*, 1714.

BRUNS, H., see Bessel.

BUCKLEY, F.
Old Watchmakers :
John Ellicott (1706–72). 8°. 6 pp., 1 pl. (P.) 1930.
George Graham (1673–1751). 8°. 6 pp., 1 pl. (P.) 1929.
Daniel Quare (1648–1724). 8°. 6 pp., 1 pl. (P.) 1930.
Old Watchmakers of London. 8°. 16 pp. (P.) 1929.
Old Watchmakers of London : Biographical Notices. 8°. 6 pp. (P.) 1930.
Old Manchester Clock and Watchmakers, 17th and 18th Centuries. 8°. 10 pp., 1 pl. (P.) 1929.

BUSCHENDORF, K. F.
Gründliche Unterricht von Thurm-Uhren, etc. 4°. *Leipsic*, 1805.

CAGNOLI, Antonio
De' Due Orologj, Italiano e Francese, ossia degl' inconvenienti che nascono del Regolare gli Orologj al Tramontar del Sole, etc. 12°. 44 pp. (P.) *Venice*, 1797.

CALAME, C.
Echappement à ancre suisse. 8 pp., 3 pl. 1882.

CALLET, F.
Essai sur les échappemens. 4°. 1778.

CAMPANI, Giuseppe
Discorso di G. C. intorno a' suoi muti oriouli, etc. *Rome*, 1660.

CAMPANI, Matth. (de Alimenis)
Horologium solo Naturæ motu atque ingenio, dimitiens & numerans momenta temporis constantissime æqualia. 4°. 27 pp., Fig. *Rome*, 1677.
(The clock is described by Allexandre, p. 321.)
The same work. 12°. (B.) *Amsterdam*, 1678.

CAMUS, Ch. E. L. de
Traité des forces mouvantes. *Paris*, 1722.
(Part II describes the construction of several watches and clocks. See also Allexandre, p. 354.)

CAMUS, Ch. E. L. de—*cont.*

Cours de Mathématique. 8°. *Paris*, (?)

Third edition, 1767. (It treats of the "numbers" of clockwheels, and is referred to by Berthoud.)

English translation, by J. Hawkins: Treatise on the Teeth of Wheels, Pinions, etc. 8°. 1806.

Second edition, 1837; Third edition. 1868.

CANCELLIERI, F.

Le due nuove campane di Campidoglio . . . con varie notizie sopra ogni sorta di orologj, etc. 4°. 216 pp., 1 pl. (P.) *Rome*, 1806.

CAPRIGLIA, Giuseppe da

Misura di Tempo, cioè Trattato d'Horologij da Ruota di tre Ordini, Rusticho da Campanile, Pulito da Camera e Lustro da Petto, etc. 4°. 72 pp., woodcuts. (B., P.) *Padua*, 1665.

(This is the *earliest work* on the construction of clocks. Although written in 1663 Capriglia only describes and illustrates balance clocks, seemingly ignorant of Huygens' pendulum clock. He does not describe the Horologio da Petto.)

CARBONNIER, E.

Neu erfundene Weckerglocke, etc. 16 figs. *Quedlinburg*, 1858.

(A translation of a French work.)

CARDANUS, Hieronymus

De Rerum Varietate. 17 vols. Fol. *Basle*, 1557.

(In Chapter 47, pp. 362–7, instructions in clockmaking are given.)

CARLIER, J. C.

Les méthodes et appareils de mesure du temps, etc. 2 vols. 8°. (S.) *Brussels*, 1905.

CARPENTER, W. B.

Mechanical Philosophy, Horology and Astronomy. 8°. (P., S.) *London*, 1857.

(Pp. 315–64 contain: Horology: or the construction of Instruments for the Measurement of Time.)

CARRÉ

Démonstrations simples et faciles de quelques propriétés qui regardent les pendules, etc. 110 pp., 1 pl. (In *Mémoires de l'Académie des Sciences.*) *Paris*, 1707.

CASPARI, Ed.

Les chronomètres de marine. 8°. 204 pp. (B., G., P.) *Paris*, [1894].

German translation: Untersuchungen über Chronometer und nautische Instrumente. 8°. 296 pp., 2 pl. By E. Gohlke. (P.) *Bautzen*, 1894.

CASSINI, Jean Dominique

Moyens de construire un pendule qui ne puisse point s'allonger par la chaleur ni se raccourcir par le froid. 4°. *Paris*, 1741.

Description of Le Roy's longitude watches. (I.) 1770.

CASSINI fils et LE ROY aîné

Voyage fait par ordre du Roi en 1768 pour éprouver les montres marines inventées par M. Le Roy. 4°. 114 pp., 11 pl. *Paris*, 1770.

CECIL, G.

Concerning Antique Clocks. (In *The Connoisseur.*) (V.) *London*, 1910–11.

CELLÉRIER, Gustave
Concours national de compensation de chronomètres pour les températures, etc. 4°. 78 pp., 12 pl. (P.) *Geneva,* 1885.
CESCINSKY, Herbert
Chapters 14–21 of "English Furniture of the 18th Century" are a Treatise on Clocks. 4°. 115 pp. *London,* 1910.
CESCINSKY, H., and WEBSTER, M. R.
English Domestic Clocks. 4°. 354 pp., very fully illustrated. *London,* 1913.
CHABERT, J. B., Marquis de
Mémoires sur l'usage des horloges marines, etc. 4°. 20 pp. (P.) (In *Mémoires de l'Académie des Sciences*) *Paris,* 1785.
CHAPUIS, A.
Histoire de la pendulerie Neuchâteloise. 4°. 502 pp., 41 pl. (B., P., V.) *Neuchâtel,* 1918.
Relations de l'horlogerie suisse avec la Chine. La Montre Chinoise. 8°. 272 pp., 33 pl. (V.) 1922.
Urbain Jürgensen et ses continuateurs. 45 pp. *Neuchâtel,* 1929.
Pendules Neuchâteloises, documents nouveaux. *Neuchâtel,* 1931.
CHAPUIS, A., and GÉLIS, E.
Le Monde des automates. 2 vols., 4°. *Paris,* 1928.
CHARLES, A. See Sully, H.
CHARMASSE, A. de
L'Horlogerie et une famille d'horlogers. 8°. (G.) *Autun,* 1888.
CHARROST
Methodical Treatise. (I.) 1795.
CHEVALLIER, L., and DUBOIS, A.
L'Emboîtage de la montre. 8°. 30 pp. (P.) *Geneva,* 1890.
CHICAGO (Town)
Watches: The Paul M. Chamberlain Collection at the Art Institute of Chicago. 8°. 58 pp. (P.) *Chicago,* 1921.
CHURCH, C. M.
The Clock and Quarter Jacks in Wells Cathedral. (In *Proceedings of the Somersetshire Archæological Society.*) (V.) 1910.
CLARKE, J.
Mercurial Chronometer Improved. (I.) 1715.
CLERKE, Gilbert
Oughtredus explicatus, ubi de constructione Horologiorum. 8°. *London,* 1682.
CLERMONT, D.
L'Horlogerie vulgarisée. 16°. 1902.
CLOCKMAKERS' COMPANY
Charter and Bye-laws. 8°. (G., P., V.) *London,* 1825.
Catalogue of the Library. (G., P.) 1898.
Some Account of the Worshipful Company of Clockmakers of the City of London, compiled by G. E. Atkins and W. H. Overall. 368 pp. (G., P.) *London,* 1881.
Catalogue of Clocks and Watches. (G.) *London,* 1875.
Second edition. 1902.

CLOCKMAKERS' COMPANY—*cont.*

A Catalogue of the Collection of the Rev. H. L. Nelthropp of Clocks and Watches, presented to the Clockmakers' Company, London. 8°. 58 pp. (G., P.) 1895.

Second edition. 1900.

CLOUZOT, H.

Les maîtres horlogers de Blois. (In the *Revue de l'Art ancien et moderne.*) (V.) *Paris*, 1913.

COLMAN, G. C.

English Bracket Clocks of the 17th and 18th Centuries. (In *The Connoisseur*, 1902.)

COOPER, T. P.

The Old Clockmakers and Watchmakers of York. (In *Associated Archæological Society's Transactions.*) (V.) 1910.

CÔTE, C.

Montres et horloges. 39 pp., 3 pl. (In *Bibliothèque des Musées de Lyon.*) (B., V.) (See also Vial.) 1919.

COULOMB, C. A.

Mémoire sur les frottements de la pointe des pivots. (In *Mémoires de l'Académie des Sciences.*) *Paris*, 1790.

COURTENVAUX, de

Journal du voyage de M. le Marquis de Courtenvaux . . . pour essayer . . . plusieurs instrumens relatifs à la longitude. *Paris*, 1768.

CRESPE, F.

Essais sur les montres à répétition, etc. 8°. 284 pp. (G.) *Geneva*, 1804.

CRISP, W. B.

Prize Essay on the Compensation Balance. 8°. 30 pp., 2 pl. (G., P.) *London*, [1875].

CUMMING, Alexander (*ca.* 1732–1814)

The Elements of Clock and Watchwork. Adapted to Practice in two Essays. 4°. 198 pp., 16 pl. (B., G., P.) *London*, 1766.

(The Library of the Clockmakers' Company possesses interleaved copies with notes and additions.)

For *German translation* see Geissler.

CUNO, Jacob

Brevis descriptio artificiosi novi et astronomici automati Horologii. 1581.

CUNYNGHAME, (Sir) H. H. S.

Time and Clocks: a Description of Ancient and Modern Methods of Measuring Time, with many Illustrations. 8°. 200 pp. (P., S.) *London*, 1906.

Theory and Practice of Clockmaking. 60 pp. (Cantor lecture, Society of Arts, vol. 8.) (P.)

DAL NEGRO

Nuovo metodo di misurare le più minute frazioni del tempo. 4°. 56 pp. *Padua*, 1816.

DALRYMPLE, A.

Instructions concerning Chronometers or Timekeepers. (I.) 1788.

DASYPODIUS, Conrad
Warhafftige Ausslegung des astronomischen Uhrwerks zu Strasburg. Descriptio Horologii astronomici Argentinensis. 4°. 1578–80.
DAVID, F. v.
Beschreibung einer astronomischen Uhr. 4°. 2 pp. *Vienna*, 1771.
DAWES, W. M.
A Familiar Treatise on Horology, etc. 8°. 30 pp. (P.) *London*, 1862.
DECRESSAIN, P.
L'Horloge électrique à l'Exposition Universelle de 1900. 8°. 116 pp., 1 pl. (P.) *Paris*, 1903.
DÉGALLIER, E.
Le Ressort de montre au point de vue de ses dimensions. 17 pp. (B.) *Bienne*, 1918.
Les Engrenages droits, etc. 8°. 176 pp., 7 pl. (P.) *Besançon*, 1923.
DELAMARCHE et PLOIX
Notes sur les marches d'un chronomètre à balancier non compensé. (*Comptes rendus de l'Académie des Sciences*, 1859.)
DEMMIN, A.
Die Uhrmacherkunst in ihrer geschichtlichen Entwickelung, etc. 8°. 48 pp., 7 illus. 1888.
DENISON, see Beckett-Denison.
DENT, Edw. J.
New Construction of the Compensation Balance. (I.) 1842.
An Abstract from Two Lectures on the Construction and Management of Chronometers, Watches and Clocks. 8°. 32 pp. (P., S.) *London*, 1842.
New edition. 36 pp. *London*, 1846.
German translation : Über die Behandlung der Chronometer, Pendel und Federuhren. *Vienna*, 1843.
On the Errors of Chronometers. 8°. *London*, 1842.
Appendix to above. 8°. 12 pp. (P., S.) *London*, 1842.
New edition of Appendix. 16 pp. (P.) *London*, 1843.
Shape of the Wheel Teeth in the Royal Exchange Clock. (I.) 1844.
DENT, F.
Treatise on Clock and Watchwork. 8°. 152 pp. (P.) *Edinburgh*, 1855.
DERHAM, William, D.D., F.R.S.
The Artificial Clockmaker. 12°. 132 pp. (G., S., V.) *London*, 1696.
Second edition, with Supplement. 12°. 119 + 28 pp. (G., P.) 1700.
Third edition, "with large emendations and additions." 12°. 140 pp. (G., P.) 1714.
Fourth edition. 12°. (G.) 1734.
Fourth edition, "corrected." 12°. 160 pp. (G., P., V.) 1759.
French translation : Traité d'horlogerie pour les montres et les pendules. 12°. (G.) *Paris*, 1731.
Second edition. 188 pp. *Paris*, 1746.
For *German translation* see Welper.
DEVELLE, E.
Les Horlogers blésois au XIVe et au XVIIe siècle. 4°. 374 pp., 18 pl. in separate vol. (B., V.) *Blois*, 1913.
Second edition, with plates interleaved. 460 pp., 30 pl. *Blois*, 1917.

Diderot et d'Alembert
Encyclopédie, see Horlogerie. *Paris*, 1751–80.

Dietzschold, C.
Die Turmuhren mit Einschluss der sogenannten Kunstuhren. 8°. 368 pp., 12 pl. (P.) *Weimar*, 1894.
Die Radverzahnungen der Uhren . . . und die Berechnung der Räderwerke. 8°. 192 pp. (P.) *Bautzen*, 1895.
Die Hemmungen der Uhren, ihre Entwickelung, Konstruction, etc. 8°. 244 pp. (P.) *Krems a/Donau*, 1905.
Abriss der Getriebelehre mit besonderer Anwendung auf die Uhrmacherei. . . . Abschnitt 2. Die Uhrgetriebe. 8°. 234 pp., 1 pl. (P.) *Krems a/Donau*, 1905.
Der Cornelius Nepos der Uhrmacher (32 portraits of important makers, etc.). 8°. (P., V.) *Krems*, 1910.
(See also Grosch.)

Dietzschold u. Zarbl
Vorlagen für das Uhrmacher Gewerbe. 4°. 177 pp., 33 pl. (P.) *Vienna and Leipsic*, 1910.
(See also Gelcich.)

Dircks, Henry
Perpetuum Mobile. (Describes James Cox's perpetual motion clock.)

Ditisheim, H.
Exécution pratique d'un échappement à ancre suisse. (Second edition.) 8°. 18 pp. (P.) *Geneva*, 1907.

Ditisheim, Paul
The timing of chronometers and watches. (Extract from *English Mechanic*.) 1926.

Doppelmayr, J. G., see Bion.

Drach, C. A. v.
Die Globusuhr Wilhelms IV von Hessen. 4°. 24 pp., 2 pl. (B.) *Marburg*, 1894.

Drexler, J.
The Drexler Simplified System of Clock and Watch Repairing. 8°. 152 pp., 49 pl. (P.) *Milwaukee*, 1914–15.

Dubois, A., see Chevallier.

Dubois, Pierre
Histoire et traité de l'horlogerie depuis son origine jusqu'à nos jours, suivie de la bibliographie des horlogers célèbres. 4°. 414 pp. (G., P.) *Paris*, 1849.
Second edition. 1850.
La Tribune chronométrique, scientifique et biographique, etc. 8°. (G.) *Paris*, 1852.
Histoire des fabriques d'horlogerie de la Suisse et de la France. 12°. *Paris*, 1853.
Collection archéologique du Prince Soltykoff (Horlogerie), etc. 4°. 214 pp., 20 pl. (P., V.) *Paris*, 1858.

Dubois-Sandoz, U.
L'Achevage de la boîte de montre. 8°. 20 pp. (P.) *Geneva*, 1901.

Duhamel, Duc
Method of ascertaining at Sea the Exact Gain and Loss of Chronometers, etc. (I.) 1832.

DUMAS, W.
Ueber Schwingungen verbundener Pendel. 8°. 14 pp. (?)
DU MONCEL
Horlogerie électrique, etc. 8°. (?)
DUMONTIER, E.
Les Bronzes du mobilier national ; pendules et cartels. (V.) 1911.
DUMONTIER, N.
L'Art de travailler les pierres précieuses à l'usage de l'horlogerie. 8°. 56 pp., 1 pl. (P.) *Paris*, 1843.
DURO, Cesâreo Fernández
Article in the *Museo Español de Antiguedades.* 1878.
(Contains history of Clockmaking in Spain and names of Makers.)
DYER, W. A.
Old Clocks. (In *Country Life.*) 1907.

EARNSHAW, T., and ARNOLD, J.
Explanation of Timekeepers constructed by the above, with a Preface by Nevil Maskelyne ; published by Order of the Commissioners of Longitude. 4°. *London*, 1806.
EARNSHAW, T.
Longitude : an Appeal to the Public, etc. 8°. 332 pp. (P.) *London*, 1806.
EBERHARDT, G. A.
Grundlinien zur Beurtheilung ganz vollkommener Thurmuhren. 8°. *Gotha*, 1812.
EDEL, F. W.
Die astronomische Münsteruhr in Strassburg. 1 pl. 1843.
EDINBURGH PHILOSOPHICAL JOURNAL
In vol. 6, April 1822, pp. 197–213, and continued in vol. 7, pp. 35–47. "History of the Invention of Pendulum Clocks by Christian Huygens," being an Abridged Translation from the Dutch of a Lecture by van Swinden (*q.v.*). (G.)
EIFFE, J. S.
Eiffe's and Molyneux's Improvements in Chronometers. (I.) 1840.
Account of Improvements in Chronometers. (Published by the Admiralty.) 4°. (G., S.) 1842.
EIGNER, P. W.
On the Balance Staff and Cylinder. 15 figs. *Chicago*. 1739.
ELLICOTT, John
Influence Two Pendulum Clocks were observed to have upon each other. (I.)
Papers communicated to the Royal Society 1739 and 1753, the latter (4°, 68 pp., 2 pl.) being A Treatise on two methods of preventing irregularities due to influence of heat and cold on pendulums. (P.)
EMELE, J.
Aufgaben-Sammlung für Uhrmacher, etc. 64 pp., 32 woodcuts. (See also Kittel.) 1891.
ENGELMANN, M.
Die Uhrensammlung Pleissner . . . zu Dresden. (In *Mitteilungen aus der sächsischen Kunstsammlungen.*) 1910.
Die Technik der Vorzeit. (Refers to the sketches of Leonardo da Vinci.) 1914.

EPHEM
Watchmakers' Almanac. 8°. *London*, 1887 *et seq.*
Watchmakers' Diary. 4°. *London*, 1891 *et seq.*

EPPNER
Entstehung der Taschenuhren-Fabrikation zu Lähn in Schlesien. 8°. 1854.

EPPS, J.
Method of ascertaining Comparative Rates of Chronometers. (I.) 1832.

ERIZZO, N.
Relazione storico-artistica della torre dell' orologio di S. Marco in Venezia, colla descrizione del mecanismo dell' orologio. (Second edition.) 8°. 190 pp., 8 pl. (P.) *Venice*, 1866.

ETCHELLS, C. T.
Repairing Repeating Watches. *Chicago.*

" EXCELSIOR," see C. E. Fritts.

EXHIBITION REPORTS
See Patent Office " Subject List of Works on Horology," pp. 16–18. *London*, 1912.

FALCONET
Dissertation sur Jacques de Dondis, auteur d'une horloge singulière ; et à cette occasion sur les anciennes horloges. Sm. 4°. (In vol. xx of *Mémoires de Littérature.*) *Paris*, 1745.

FALLET-SCHEURER, M.
Geschichte der Uhrmacherkunst in Basel 1370–1874. 284 pp. (B.) *Berne*, 1917.

FARMER, J. J.
An Account of the Improvement in Chronometers, Watches and Clocks. (A Second edition of Loseby's Work, *q.v.*) *London*, 1914.

FAVARGER, A.
L'Electricité et ses applications à la chronométrie. 8°. 133 pp., 104 figs. *Geneva*, 1886.
Second edition. 8°. 198 pp., 139 figs. (B.) *Geneva*, 1892.
Third edition. 8°. 558 pp., 8 pl. *Neuchâtel*, 1924.
German translation of Second edition, by M. Loeske : Die Elektrizität und ihre Verwerthung zur Zeitmessung. 8°. 243 pp., 139 illus. *Bautzen*, 1895.

FAVRE, M. et Cie
Echappements fixes, carrousels et tourbillons. Second edition. 8°. 16 pp. (P.) *Geneva*, 1908.

FERGUSON, J. (F.R.S.)
Tables and Tracts Relative to Several Arts and Sciences. (This includes horology.) 8°. (G.) *London*, 1767.
Select Mechanical Exercises : showing how to construct Different Clocks, etc. 8°. 324 pp., 9 pl. (P.) *London*, 1773.
Second edition, 1778 ; Third, 1790 ; Fourth, 1823.

FERRET, E.
Les Breguets. 8°. 132 pp. (P.) *Paris*, 188?.

FÉTIL, l'aîné
La Théorie de l'horlogerie. 8°. *Paris*, An XI.

FICHOT, E., et VANSSAY, P. de
Congrès international de chronométrie : Comptes rendus des travaux. 4°. (S.) *Paris*, 1902.

FIEDLER, L.
Die Zeittelegraphen und die elektrischen Uhren, etc. 8°. 207 pp., 85 figs. (B.) *Vienna*, 1890.

FINÉE, O.
Description de l'horloge planétaire du Cardinal de Lorraine, de l'invention d'Oronce Finée. 4°.

FINN, J. L.
Poising the Balance. *Chicago*.

FINN, J. L., and RIEFLER, S.
Compensating Pendulums and how to make them. *Chicago*, 1894.

FISCHER, F.
Anleitung zur Berechnung der Räder an Uhren. 8°. *Villingen*, 1841.

FISHER, Rev. G.
On the Errors in Longitude as determined by Chronometers at Sea arising from the action of Iron in the Ships, etc. (In *Philosophical Transactions*.) 4°. 16 pp., 5 tables. (P.) *London*, 1820.

FLEURIEN, d'E. de
Voyage fait par l'ordre du Roi en 1768 et 1769 à différentes parties du monde pour éprouver en mer les horloges marines inventées par F. Berthoud. 4°. 2 vols. *Paris*, 1773.

FLEURY, Paul de
Note sur une horloge à pendule régulateur construite à Angoulême quatorze ans avant la naissance d'Huygens. (In *Bulletin de la Société archéologique et historique de la Charente*.) *Angoulême*, 1890–1.

FÖRSTER, W.
Ueber Zeitmessung und Zeitregelung. 8°. 114 pp. (P.) *Leipsic*, 1909.

FORSTMANN, C. W.
Ausführliche Unterricht von zeigenden und schlagenden Taschenuhren, etc. 8°. 562 pp., 10 pl. (P.) *Halle*, 1779.

FOUCHER, P.
Manuel d'horlogerie. 12°. 1867.
Italian translation: Manuale de Orologeria. 1885.

FOUCHY, J. de
Manière fort simple de se servir d'horloges de moyen volume, etc. *Paris*, [1740].

FRANCE HORLOGÈRE (La)
Recueil de procédés pratiques de rhabillage d'horlogerie. *Besançon*.
Second edition. 8°. 212 pp. (P.) *Besançon*, 1904.

FRANKLIN, A.
La mesure du temps. 12°. 239 pp. (P.) *Paris*, 1888.
Second edition. 1896.

FRAUBERGER, H.
N. P. Fraenkel's Uhrensammlung : herausgegeben von H. Frauenberger. Fol. (B., V.) *Düsseldorf*, 1913.
FRÉMONT, Charles
Origine de l'horloge à poids. 4°. (P.) *Paris*, 1915.
Origine de la Poulie, du Treuil, de l'Engrenage, etc. 4°. *Paris*, 1921.
FRIES, see Ludwig und Fries.
FRITTS, C. E. (" Excelsior ")
A Practical Treatise on the Balance Spring, etc. 8°. 144 pp. (P.) *New York*, 1876.
The Watch Adjuster's Manual. 8°. 376 pp. (P.) *London and New York*, 1894.
Fourth edition. 8°. 366 pp. *Philadelphia*, 1912.
FRODSHAM, C.
Vibrations of Pendulums with different Suspension Springs. (I.) 1838.
A Few Facts connected with the Elements of Watch and Clockmaking. *London*.
On the Laws of Isochronism of the Balance Spring, etc. 8°. 34 pp. (P.) *London*, 1849.
On Mercurial Compensation Pendulums, etc. 8°. 8 pp. (P.) *London*, [1866].
(See also Parkinson.)
FROISSART, Jean (1338–1410)
" Li Orloge amoureus." (In vol. i of Scheler's edition of the Poems.) *Brussels*, 1870.
Chronicles. (Vol. ii., chap. 128, describes the Duke of Burgundy's clock.)
FROMBERG, E. O.
Das Abziehen, Ausputzen, Repariren und Reguliren . . . der Taschenuhren, etc. 8°. 66 pp., 1 pl., illus. *Quedlinburg*, 1850.

GABORY, M.
Manuel utile et curieux sur la mesure du temps, contenant des méthodes pour régler les montres et les pendules (pp. 1–54). 12°. 135 pp. (P.) *Angers*, 1771.
GAGNANT, L.
Eléments de calcul des rouages des montres et des pendules. 8°. 64 pp. (P.) (?)
Etude sur le pendule. 8°. 24 pp. (P.) *Geneva*, 1906.
Tracé des échappements de pendule les plus employés, etc. 8°. 16 pp. (P.) *Besançon*, 1909.
La Pendule de Morez. 8°. 86 pp. (P.) *Besançon*, 1912.
GAMMENTHALER, A.
L'Achevage de l'échappement à ancre après dorure. 8°. 20 pp. (P.) *Geneva*, 1911.
Le Remontage du rouage. 8°. 20 pp. (P.) *Geneva*, 1912.
GARDNER, D.
The Ship's Chronometer : its History and Development. 26 pp., 6 illus. 1890.

Garnier, Paul
Rapport sur l'horloge à l'Exposition Universelle de 1889 à Paris. (V.)
Garrard, F. J.
Watch Repairing, Cleaning and Adjusting. 8°. 224 pp. (P.) *London*, 1903.
Second edition, 1908. (P., S.) Fifth edition, 1919. (P.)
Clock Repairing and Making. 8°. 176 pp. (P., S.) *London*, 1914.
Garuffa, E.
Orologeria moderna. 8°. 302 pp. (B., P.) *Milan*, 1894.
Gay-Lussac, L. J., and Arago.
Breguet's Chronometers. (I.) 1819.
Geiser
Ueber die Geisersche sich selbst im Gange erhaltende Pendeluhr. 12°. 80 pp. *Frankfort*, 1818.
Geissler, J. G.
Der Uhrmacher, oder Lehrbegriff der Uhrmacherkunst, etc. 10 parts. 4°. 87 pl. (P.) *Leipsic*, 1793–9.
Elemente der praktischen Gross- und Kleinuhrmacherkunst. 8°. 192 pp., 16 pl. (P.) *Leipsic*, 1802.
(This is a *German translation* of Alex. Cumming's work, *q.v.*)
Gelcich, Eugen
Geschichte der Uhrmacherkunst von den ältesten Zeiten bis auf unsere Tage. 8°. 208 pp., 4° atlas. (B., P.) *Weimar*, 1887.
(The Fourth edition of Barfuss' work, *q.v.*)
Die Uhrmacherkunst und die Behandlung der Präzisionuhren. 8°. 656 pp., 249 illus. (B., P.) *Vienna*, 1892.
Skizze einer Geschichte der Chronometer, etc. (In the *Uhrmacher Zeitung*.) *Berlin*, 1885–6.
Gelcich, E., und Dietzschold, C.
Die Tabellen der Uhrmacherkunst, nebst einer Sammlung mathematischer Hilfstafel für Uhrmacher. 8°. (B.) *Vienna*, 1892.
Die Räderverzahnungen der Uhren, etc. 8°. 190 pp. *Bautzen*, 1895.
Gélis, E., see Chapuis.
Georges, Pierre
Horloge magnétique . . . pour trouver les heures du jour et de la nuit. 8°. *Toul*, 1660.
Georgi, Erasmus
Handbuch der Uhrmacherkunst. 8°. *Altona*, 1867.
Second edition. 618 pp., 232 woodcuts. 1883.
Gerland, E.
Zur Geschichte der Erfindung der Pendeluhr . (In *Wiedemann's Annalen der Physik u. Chemie*, vol. iv, pp. 585–613.) *Leipsic*, 1878.
Die Erfindung der Pendeluhr. (In *Westermann's Monatshefte* for August 1884.) (B.)
(He ascribes the invention of the pendulum to Huygens, and describes and illustrates the Dover Castle Clock.)
Gfeller, J.
L'Horlogerie suisse en 1886. *Bienne*, 1886.

GIEBEL, K.
Der Einfluss der Hemmung auf den Gang der Uhr. 8°. 58 pp. *Berlin*, 1906.

GIRAULT, C.
De la résistance de l'air dans le mouvement du pendule. 8°. 46 pp. *Caen*, 1858.

GIVRY
Mémoire sur l'emploi des chronomètres à mer. 135 pp. *Paris*, 1840.

GLASGOW, D.
Watch and Clockmaking. 8°. 354 pp. (G., P., S.) *London*, 1885.
Second edition. 1897.

GLORIA, A.
L'Orologio di Jacopo Dondi. *Padua*, 1885.
I due orologi meravigliosi inventati da Jacopo e Giovanni Dondi. (In vol. 54, pp. 675–736, of the *Atti del Reale Istituto Veneto di Scienze*, etc.) 8°. 1896.

GOHLKE, E., see Caspari.

GOLFARELLI, I.
Cenni storici dell' Orologeria. 8°. 46 pp. (B.) *Florence*, 1871.
L'Orologio misterioso inventato dal Signor Guilmet. 8°. 8 pp., 1 pl. (B.) *Florence*, 1875.

GONDY, J.
Manuel d'horlogerie. 12°. 388 pp. (B., P.) *Paris*, 1910.

GOODCHILD, J. G.
On the Evolution of the Japanese Clock. (In *Transactions of the Royal Scottish Society of Arts*, vol. 16.)

GOODISON, I. C.
The Article on (English) Clocks in the *Dictionary of English Furniture*, vol. ii. *London*, 1924.

GOODRICH, W. L.
The Watchmaker's Lathe, its Use and Abuse. 8°. 267 pp. *Chicago*, 1903.
The Modern Clock. *Chicago*, 1905.

GORDON, G. F. C.
Clockmaking, Past and Present. 8°. 240 pp., 35 pl. (P.) *London*, 1925. (Incorporating portions of Beckett-Denison's "Clocks and Watches and Bells.")

GOULD, Lt.-Comm. R. T.
The Marine Chronometer : its History and Development. 8°. 303 pp., 62 pl. (P.) (An exhaustive work on the subject.) *London*, 1923.

GRAFFIGNY, H. de
Les merveilles de l'horlogerie. 8°. (B.) 1888.
Manuel de l'horloger et du mécanicien amateur, etc. 12°. 336 pp., 230 figs. (B., P.) *Paris*, [1894].

GRAFTON, E.
Horology : or a Popular Sketch of Clock and Watchmaking. Third edition. 74 pp. (P.) *London*, [1849].

GRAHAM, George.
Contrivance to avoid the Irregularities in a Clock's Motion. (I.) 1726.

GRALL, G., see Bourdais.

GRAMETBAUER, H.
Neues Gewicht-System für Gewichtuhren, etc. 1864.

GRAUPMANN, E.
Le finissage de la boîte de montre. 8°. 20 pp. (P.) *Geneva*, 1910.

GREEN, F. H.
Old English Clocks, etc. (Full plates of works of the great masters.) 1931.

GRENIER.
Note on Astronomical Clocks. (I.) 1790.

GRIBI, Theo
Practical Course in Adjusting. 8°. 212 pp. *New York*, 1901.
Cours pratique de réglage. 8°. 150 pp. (P.) *Geneva*, 1910.
(See also Schouffelberger.)

GRIFFITH, W. F. and J. W.
Revised and enlarged edition of Johnston's translation from German of Beckman's History of Inventions. (B.) *London*, 1846.

GRIMTHORPE, Lord, see Beckett-Denison.

GROS, C.
Echappements d'horloges et de montres. 125 pp. *Paris*, 1913.
(See also Schultz, W.)

GROSCH, H., DIETZSCHOLD und HÜTTIG
Praktisches Handbuch für Uhrmacher. 8°. 279 pp. and atlas. (P.) *Weimar*, 1879.
Second edition. 8°. 328 pp. (P.) *Leipsic*, 1907.

GROSCLAUDE, L. A.
L'échappement à ancre, etc. 8°. 36 pp., 5 pl. *Geneva*, 1884.

GROSSMANN, J.
Lehrbuch der Uhrmacherei nach den Gesetzen der Mechanik. (*German translation* of Arndt and Defossez.) 8°. 258 pp., 2 pl. (P.) *Bautzen*, 1902– .

GROSSMANN, J. and H.
Lessons in Horology. (Translation by J. Allan.) 8°. (P.) *Philadelphia*, 1905– .
Horlogerie théorique. 2 vols. 8°. (P., S.) *Bienne*, 1911–12.

GROSSMANN, M., see Saunier.

GROSSMANN, Moritz
Der freie Ankergang für Uhren. (Prize Essay.) 8°. 207 pp., atlas. *Glashütte*, 1866.
Second edition, by L. Strasser. 8° atlas. (P.) *Bautzen*, 1892.
Third edition. 1896.
The Detached Lever Escapement. (Author's English edition.) 188 pp. 1866.
English translation : A Practical and Theoretical Treatise on the Detached Lever Escapement. 8°. 212 pp. *Chicago*, 1884.
French translation : L'Echappement libre à ancre. 8°. 20 pl. 1867.
Taschen Wörterbuch für Uhrmacher in drei Sprachen Deutsch, Englisch, Französisch. *Glashütte*, 1880.
Second edition. *Bautzen*, 1891.

GROSSMANN, Moritz—*cont.*

Horological Pocket Dictionary in Three Languages, English, German and French. In 3 vols. *Glashütte*, 1880.

Second edition. *Bautzen*, 1891.

A Practical and Theoretical Treatise on the Detached Escapement for Watches and Timepieces. 8°. 188 pp., 20 pl. (P.) *Leipsic*, 1866.

Abhandlung über die Konstruktion einer einfachen aber mechanisch vollkommenen Uhr. 8°. 135 pp., 58 illus. *Glashütte*, 1880.

Second edition. 1890.

Third edition. 98 pp. (P.) *Bautzen*, 1897.

English translation: "On the Construction of a Simple and Mechanically Perfect Watch." (Prize Essay.) 8°. 96 pp. (P.) *Chicago*, 1891.

Das Regulieren der Uhren für den bürgerlichen Gebrauch mit Tabellen. 8°. 60 pp., 14 illus. (Second edition.) *Bautzen*, 1890.

Third edition. 8°. 58 pp. (P.) *Bautzen*, 1897.

(See also Lossier and Saunier).

GUILDFORD, E. L.

Memorials of Old Nottinghamshire.

(Gives list of Newark-on-Trent Clockmakers.)

GUILLAUME, C. E.

Calcul d'un pendule compensé en acier-nickel. 8°. 11 pp. (B.) *Geneva*, 1897.

Le pendule en acier au nickel, etc.

Second edition. 8°. 60 pp. (P.) *Geneva*, 1908.

GUMBEL, A.

Peter Henlein, der Erfinder der Taschenuhren. 8°. (P.) *Halle*, 1924.

HAËFTEN (Benedictine)

Monasticæ Disquisitiones. Fol. *Antwerp*, 1644.

(Book vii, disquisition 2, tells how the monks in early times measured the hours by day and night.)

HAHN, J. G.

Ueber Verbesserungen von Taschenuhren, etc. 4°. 1 pl. *Erfurt*, 1784.

HAINAUT, R. L.

Etude historique des progrès de l'échappement à cylindre. Nouveau cylindre incassable. *Rouen*, 1885.

Recherches sur l'origine de l'horlogerie. 8°. 118 pp. 1 pl. (P.) *Rouen*, 1896.

HAINDL

Konstruktion der Verzahnungen, etc. 8°. *Stuttgart*, 1830.

HALL, E. E.

A Treatise on Staff-making and Pivoting. 48 pp., 24 figs. *Chicago*.

HALL, Wm.

A New and True Way to find the Longitude. *London*, 1714.

HALLE, J. S.

Der Uhrmacher. *Brandenburg*, 1761.

HALLWARD, Reginald

Some Surviving Handicrafts. (In the *Architectural Review*, 1900.)

HAMBERGER, see Beckmann.

Hammer, R.
Nürnberg : Historische Uhrenausstellung. 8°. 108 pp., map. (P.) *Nuremberg*, 1905.

Hanke
Uhrmacherlehre. *Leipsic*, 1912.

Hansen, F., see Wiedemann.

Hansen, P. H.
Ueber die Chronometer welche Herr Kessels verfertigt, etc. 8°. 55 pp., 33 pl. 1836.

Harrison, John
Portsmouth to Port Royal : Experiment with Harrison's Timekeeper. (I.) 1762.
An Account of the Proceedings in order to the Discovery of the Longitude at Sea ; relating principally to the Time-piece of Mr. John Harrison. . . . By a member of the Royal Society. 8°. 98 pp. (P.) *London*, 1763.
A Narrative of the Proceedings relative to the Discovery of Longitude at Sea by Mr. John Harrison's Time-keeper subsequent to those related in 1763. (P.) *London*, 1765.
The Principles of Mr. Harrison's Timekeeper, by the Rev. N. Maskelyne, published by Order of the Commissioners of Longitude. 4°. 32 pp. (G., P.) *London*, 1767.
French translation : Principes de la Montre de M. Harrison. By Père Pézénas. 4°. *Avignon*, 1768.
Remarks on a Pamphlet lately published by the Rev. Mr. Maskelyne. 8°. 34 pp. (G., P.) *London*, 1767.

Harrison, Thomas and James
The Complete Observator, containing the Various Methods of finding the Longitude . . . and of correcting the Watch. 4°. *York*, 1788.

Hartmann, J. G.
Unterricht von Verbesserung der Sack-Uhren, wie auch Erklärung der üblichen Kunstwörter. 8°. 1 pl. *Jena*, 1752.
Second edition enlarged. 8°. 120 pp., 1 pl. (G., P.) *Halle*, 1756.
Nöthiger Unterricht von Verbesserung aller Uhren, etc. 8°. 246 pp., 16 pl., 2 tables, portrait. (G., P.) *Halle*, 1752.
Second edition. (G., P.) *Halle*, 1756.
Le temps vrai et le temps moyen, etc. (P.) *Paris*, 1827.

Hasluck, P. N.
Watch Jobber's Handybook, etc. 8°. 144 pp. (P., S.) *London*, 1887.
(Pages 50–140 give a glossary of terms.)
Second edition, 1889. Third edition. (S.) 1893.
American Edition. *New York*, 1902.
The Clock Jobber's Handybook, etc. 8°. 159 pp. (P.) *London*, 1889.
Second edition. (S.) 1893.

Haswell, J. E.
Horology : the Science of Time Measurement, and the construction of Clocks, Watches and Chronometers. 8°. 283 pp., illus. (P.) 1928.

Hatton, Thos.
Introduction to the Mechanical Part of Clock and Watchwork. 8°. 400 pp., 18 pl. (G., P.) *London*, 1773.

HAUTEFEUILLE, Abbé Jean de
Factum de M. l'Abbé de Hautefeuille touchant les pendules de poches, contre M. Huygens. 4°. 19 pp. (B.) *Paris*, 1675.
Avis sur le privilège des horloges et les montres de la nouvelle invention.
La pendule perpetuelle. 4°. *Paris*, 1678.
Inventions nouvelles. 4°. *Paris*, 1717.
Construction nouvelle de trois montres portatives, etc. 4°. *Paris*, 1722.

HAVARD, H.
L'Horlogerie. 8°. 182 pp. (B.) *Paris*, 1893.
(Also Articles on Clocks in "*Les Arts de l'ameublement*," and "*Dictionnaire de l'ameublement . . . depuis le* 13e *siècle*.")

HAWKINS, J., see Camus.

HAYDEN, Arthur
Chats on Old Clocks. 8°. 302 pp. *London*, 1917.

HEARN, Mrs. G. A.
Collection of Watches loaned to the Metropolitan Museum of Art. 8°. 29 pp., 9 pl. *New York*

HEATON, E.
A New Course about Modern Watchmaking, in English and French. Third edition. (P.) *Chaux-le-Fonds*, 1917.
Terminologie horlogère française et anglaise. 75 pp. (B.) *Bienne*, 1918.
A translation of French horological terms into English equivalents.

HEIDNER, G.
Die Schule des Uhrmachers. Illus. *Wiesbaden*, 1865.
Second edition. 8°. 182 pp. 1871.

HEINE, Xavier
Allgemeine Grundsätze über die Uhrmacherei, etc. 8°. *Villingen*, 1849.

HELFENZIERDER, J.
Beiträge zur Verbesserung der Uhrmacherkunst in Rücksicht auf grosse Uhren. *Munich*, 1789.
Fortsetzung. *Munich*, 1797.

HELWIG, A.
Drehganguhren. (Tourbillons und Karusselluhren. Ihr Bau, ihre Eigenarten; Vorzüge und Nachteile der verschiedenen Konstruktionen). 8°. 100 pp. (P.) *Berlin*, 1927.

HENDERSON, Ebenezer
An Historical Treatise on Horology, shewing its rise and progress from the earliest period down to the present time. 8°. 24 pp. (B., G., P.) *London*, 1836.

HERNE
Longitude Unveiled. 1678.

HERRMANN, Fr.
Katechismus der Uhrmacherkunst, etc. 8°. 100 pp. 57 illus. *Leipsic*, 1863.
Second edition. 1874.
Third edition. 1884.
Fourth edition. (B.) (See Rüffert.) 1885.

HERTZ, G., see also Schultz, W.
Geschichte der Uhren. 8°. 66 pp., 19 illus. *Berlin*, 1851.

HESSEN
Mémoire sur l'horlogerie, etc. 8°. 44 pp. (P.) *London*, 1785.

HESSLING, E.
Documents de style Empire ; Orfévrerie et horlogerie. (V.) 1913.

HEVELIUS (Johann Hevel)
Machina Cœlestis, 1673. (P. 366 refers to Huygens' invention in 1657, also to clocks of his own invention.)

HEVESI, L.
Eine Ausstellung alter Fächer und Uhren in Wien. (In "*Kunst u. Kunsthandwerk.*") (V.) 1903.

HIGGINBOTHAM, C. T.
The Watch Balance and its Jeweling. 16°. 15 pp. 1912.
Precision Time Measurers, their Construction and Repair. 345 pp., many illus. *Chicago*, 1913.

HILDEYARD, Prof. Thomas
Chronometrum mirabile Leodiense. 4°. (G.) *London*, 1727.

HILFIKER
L'Influence de la pression de l'air sur la marche des chronomètres. *Neuchâtel*, 1889.

HILLMANN, B.
Der Kronenaufzug. 8°. 48 pp. (P.) *Halle*, 1910.
Die Reparatur komplizierter Taschenuhren. 8°. 120 pp. 5 pl. (P.) *Berlin*, 1911.

HIPP, M.
Notice relating to the Electric Clocks manufactured by M. H. 12°. 24 pp. 2 pl. (P.) *Neuchâtel*, 1876.

HIRE, de la
Remarques sur la construction des horloges à pendule. 1700.
Remarques sur les inégalités du mouvement des horloges à pendule. 4°. 14 pp. *Paris*, 1703.
Sur les pendules à secondes. *Paris*, 1715.
Construction d'une horloge qui marque le temps vrai avec le temps moyen. *Paris*, 1717.
(The above articles are contained in the *Mémoires de l'Académie Royale des Sciences.*)

HIRN, G. A.
Théorie et application des pendules à deux branches. 14 pp. 1887.

HIRSCH, Dr.
La pendule électrique de précision de M. Hipp. 1884.

HOBBS, Wm.
A New Discovery for Finding the Longitude. *London*, 1714.

HOEFER, F. A.
Geschiedenis der openbare Tijdaanwijsing. *Leiden*, 1887.
Bijdrage tot de Geschiedenis der Tijdmeters. (In the *Tijdschrift voor Geschiedenis.*) *Leiden*, 1889.

HOLDER, Wm., D.D.
A Discourse concerning Time. 12°. (B.) *London*, 1694.
Second edition. (B.) 1701.
Third edition. (B.) 1712.

HOMANNS, J. B.
Neu inventirte geographische universal Zeig- und Schlag-Uhr, welche verfertigt Zach. Landteck, in Nürnberg. Copper-plate. *ca.* 1720.

HOOD, G.
Modern Methods in Horology. 8°. 253 pp., 5 pl. (P.) *Kansas City*, 1904.

HOPKINS, C.
Watchmaker's and Jeweler's Handbook. 8°. 40 pp. *Louisville*, 1866.

HORLOGERIE, L' Art de l', enseigné en 30 leçons, ou manuel complet de l'horloger et de l'Amateur . . . par un élève de Breguet. 2 vols. 12°. 17 pl.
(See Anonymous Works.)

HORLOGERIE
Important article in Diderot et d'Alembert's *Encyclopédie des Sciences, des Arts et Métiers*. Fol. 64 pl. *Paris*, 1765.

HOROLOGICAL JOURNAL, p. 64
English Clocks and Watches and their Makers. [1924.]

HOROLOGICAL MISCELLANY
(Description of Clocks and list of works) *London*, 1821.

HORRMANN, H.
Praktische Abhandlung über die Repassage einer viersteinigen Cylinderuhr. 8°. 94 pp. *Leipsic*, 1881.
Second enlarged edition. 8°. 150 pp., 35 illus. *Halle*, 1888.

HORSTMANN, G. H.
Taschenuhren früherer Jahrhunderte aus der Sammlung Marfels. 8°. 14 pp., 24 pl. (B., G.) *Berlin*, 1897.

HOSOKAWA, Hanzo
He wrote a book in 3 volumes in 1796, of which one is entirely devoted to clocks, with 18 illustrations of all the parts used in making a Japanese single-balance "lantern" clock and the completed mechanism.

HOUDIN, R.
L'Art de connaître et régler les montres et les pendules. *Paris*, 1863.
German translation : Die Kunst Taschen- und Pendeluhren nach ihrem Werthe zu schätzen, sie im Gange zu erhalten, und zu regulieren. 16°. 60 pp. *Zurich*, 1865.

HOURIET, Ch.
Les fabriques américaines d'horlogerie et leur outillage. (In the *Journal suisse d'horlogerie*.) 45 pp., 47 illus. *Geneva*, 1895.

HOURIET DU LOCHE, F.
Isochronism of Spiral Springs to render Clocks more Exact. (I.) 1814.

HOWGRAVE-GRAHAM, R. P.
Peter Lightfoot, Monk of Glastonbury, and the Old Clock at Wells. 8°. 56 pp., 8 pl. (P., V.) *Glastonbury*, 1922.
Some Clocks and Jacks, with Notes on the History of Horology. 54 pp. (Article in *Archæologia* for 1928. It deals with the clocks of Dover, Wells, Exeter, Wimborne, St. Mary Ottery, etc.)

Huet, F.
Dissertation sur l'horlogerie. 8°. 18 pp. (P.) *Paris*, 1791.
Hugon
Etude théorique et pratique sur les engrenages. 8°. Illus.
Huguenin, I.
Etablissage et repassage des montres à répétition. 8°. 31 pp. (P.) *Geneva*, 1883.
Second edition enlarged. 8°. 48 pp. (P.) *Geneva*, 1897.
Hupp, O.
Regensburger Uhren im Rathhaus zu Regensburg. (V.) *Ratisbon*, 1910.
Hüttig, see Grosch.
Hux, J. R.
A Letter on Horology. 4°. 12 pp. (P.) *London*, 1862.
Huygens, Christiaan
Horologium. 4°. 12 pp., 1 pl. *Leiden*, 1658.
French translation of "Horologium," by Léopold Reverchon, in *L'Horloger* for November 1922. (It contains a few inaccuracies.)
Horologium Oscillatorium, sive de motu pendulorum ad horologia aptato demonstrationes geometricæ. 4°. 176 pp. (B., G., P.) *Paris*, 1673.
Die Pendeluhr. *German translation*, by A. Heckseher, in Ostwald's "Klassiker." 8°. 268 pp. (B., P.) *Leipsic*, 1913.
Brevis institutio de usu horologiorum ad inveniendas longitudines. 16 pp.
Opera varia, 2 vols. : 's Gravesande's edition. (Vol. I contains the above works.) *Leiden*, 1724.
Journal des Sçavans, of February 25, 1675, contains Huygens' explanation of his invention of the spiral spring.

Immisch, M.
Der Isochronismus der Spiralfeder, etc. 8°. *Weimar*, 1871.
Second edition. 8°. 56 pp. (P.) *Weimar*, 1879.
Prize Essay on the balance spring and its isochronal adjustments. 8°. 50 pp. (P.) *Chicago*, 1872.
French translation of the Prize Essay, by A. Buisson : Mémoire sur le spiral et son isochronisme. 8°. (In the *Revue chronométrique*.) *Paris*, 1876.

Jacquemart, A.
Histoire du mobilier. (Notes on Clocks and Watches.) 1876.
Jahn, A.
Anleitung zur genauen Bestimmung des Ganges und Standes der Uhren, etc. 8°. 240 pp., 11 tables, 2 pl. 1842.
James, E.
Nouvelle méthode facile et exacte pour déterminer rapidement les dimensions des engrenages. Second edition. *Geneva*, 1897.
Cours pratique de réglage et de précision. 8°. 1898.
Traité des sonneries, etc. 12°. 100 pp. (P.) *Geneva*, 1899.
Théorie et pratique de l'horlogerie, etc. 16°. 278 pp. *Paris*, 1906.

JANVIER, Antide

Exposition publique, etc. 4°. 16 pp. (P.) *Paris*, 1802.

Etrennes chronométriques. *Paris*, 1810.

Essai sur les horloges publiques pour les communes de la campagne. 8°. 62 pp., 3 pl. (G., P.) *Paris*, 1811.

Des révolutions des corps célestes par le mécanisme des rouages. 4°. (G.) *Paris*, 1812.

Manuel chronométrique. *Paris*, 1815.

Second edition. 12°. 278 pp., 5 pl. (G., P.) *Paris*, 1821.

Recueil de machines composées et exécutées par A. J., etc. 4°. 40 pp., 12 pl. (P.) 1827.

JANVIER, L., see Berthoud, F.

JANVIER et MAGNIER

Handbuch der Uhrmacherkunst (nach den französischen Werken von Janvier und M. D. Magnier). 2 Parts. 8°. 14 pl. *Leipsic*, 1851.

Part II, by P. J. Krüger (*q.v.*). *Berlin.*

JENKINS, H.

A Description of several Astronomical and Geographical Clocks, etc. *London*, 1760.

Second edition. 8°. (G., V.) *London*, 1778.

JEROME, C.

History of the American Clock Business for the Past Sixty Years. 12°. 152 pp. *New Haven*, 1860.

JEWELER'S CIRCULAR

Workshop Notes for Jewelers and Watchmakers. 4°. 200 pp. (P.) *New York*, 1892.

JIŘÍK, F. X.

Katalog der Uhren-Ausstellung. (Held in Prague, April 5 to May 17, 1914.) Sm. 4°. 47 pp. *Prag*, 1914.

(It contains a list of Bohemian makers.)

JODIN, Jean

Traité des échappemens, etc. 1754.

Les échappemens à repos comparés aux échappemens à recul. 12°. 250 pp., 3 pl. (G.) *Paris*, 1762.

JOHANN, A.

Allgemeines Lehrbuch der Uhrmacherkunst. 8°. 325 pp., 35 pl. *Aarau*, 1879.

French translation : Traité général d'horlogerie. 36 pl.

JOHNSON, E. D.

Improvements in Chronometers. 1842.

JONES, B. E.

Clock Cleaning and Repairing. 8°. 156 pp. *London*, 1917.

Watch Cleaning and Repairing. 8°. 156 pp. *London*, 1918.

(Both handbooks, fully illustrated, were based on contributions to *Work*, by " G.")

JONES, F. Hope

Electric Clocks. 8°. 240 pp., 126 diagrams. *London*, 1931.

JONES, J.

A Sketch of the History and Principles of Watchwork. 8°. 16 pp., 4 pl. (P.) 1845.

JOSEPHI, W.
Künstlerische Taschenuhren. *Brunswick*, 1906.
JÜRGENSEN, Urban (1776–1830)
Principes généraux de l'exacte mesure du temps par les horloges. *French translation* by L. U. Jürgensen. 4°. 288 pp., 19 pl. (G., P.) *Copenhagen*, 1805.
Second edition (with a bibliography). 4°. 278 pp. (B., P.) *Paris*, 1838.
Third edition. 1865.
German translation : Allgemeine Grundsätze der genauen Zeitmessung durch Uhren. 4°. *Copenhagen*, 1806.
Mémoires sur l'horlogerie exacte : translated from Danish by his son, L. U. Jürgensen. 4°. (B., G., P.) *Paris*, 1832.
German translation, by L. U. J. : Die höhere Uhrmacherkunst. 8°. (B.) *Hamburg*, 1842.
English translation : The Higher Horological Art.
Regler for tidens nøiagtige afmaaling . . . ved uhre, etc. (Second edition by L. U. J.) 4°. 226 pp., atlas. (B., P.) *Copenhagen*, 1839.
JÜRGENSEN, L. U. (1806–1867)
Uhrmagerkunstens literatur . . . samlede i chronologisk orden. (Bibliography.) 4°. 628 pp. (P.) *Copenhagen*, 1839.
Fortegnelse over de Arbeider, som udføres i U. Jürgensen's sønner's Chronometer- og Uhr-Etablissement. 4°. 16 pp., 1 pl. (B., P.) (Factory catalogue of U. Jürgensen's sons.) *Copenhagen*, 1836.
Beretning om chronometre af in nyere construction of Urban Jürgensen's sønner's etablissement. 4°. 26 pp. (P.) *Copenhagen*, (?) 1845.
Continuations-Beretning over et chronometer af simplificeret construction forfaerdiget af Urban Jürgensen's sønner. 8°. 8 pp. (P.) *Copenhagen*, 1858.
Sur le mouvement du pendule. *Copenhagen*, 1853.

KAISER, F. (Astronomer)
De gang van het sterrekundige Slingeruurwerk, etc. (With other treatises.) *Leiden*, 1860.
KARMARSCH, K.
Versuche über die Abnahme der Zugkraft einer in einem Federhause eingeschlossenen sich abwindenden Uhrfeder. (In vol. 18 of "Jahrbücher des K.K. polytechnischen Institutes in Wien.") *Vienna*, 1834.
KASTNER, A. G.
Über die Aenderung des Ganges der Pendeluhren im Sommer und Winter. *Göttingen*, 1778.
Der neue englische Uhrmacher. 1781.
KEELHOFF, F.
La suspension à ressort. 8°. 32 pp. (P.) *Geneva*, 1909.
KEMLO, F.
Kemlo's Watch Repairer's Guide. (V.) 1869.
Kemlo's Watch Repairer's Handbook. 12°. 94 pp. (P.) *Philadelphia*, 1877.
New edition. 8°. (G.) *Philadelphia*, 1882.
The same. 1918.

KENDAL, J. F.
A History of Watches and other Timekeepers. 8°. 247 pp. (G., P., S.) *London*, 1892.

" KEYSTONE "
Watch and Clock Escapements. 8°. 180 pp. (P.) *Philadelphia*, 1904.

KIENLE, H.
Untersuchungen über Pendeluhren, mit besonderer Berücksichtigung der beiden luftdichten Riefler Uhren . . . der k. Sternwarte zu München. 4°. 103 pp. (B.) *Munich*, 1918.

KINDLER, P. F.
Die Uhren, etc. 8°. 190 pp. 1905.

" KINOSTAN "
Electrical Clocks : Principles, Construction and Working. 8°. 152 pp. (P.) *London*, [1920].

KITE, E.
Old Wiltshire Clocks and Clockmakers. (In *Wiltshire Notes and Queries*, 1909.) (V.)

KITTEL, A.
Die Rad-Verzahnungen, etc. 8°. 42 pp., 21 illus. 1889.
Lehrbuch für den Uhrmacher in Schule und Werkstatt. 163 illus. 1893.
Der Konstrukteur und das Konstruktionszeichnen in der Uhrmacherei. (In Part III of the *Bibliothek der Handelszeitung*.) *Leipsic*.
Konstruktions- und Lehrbuch der Uhrmacherei. 8°. 102 pp. *Leipsic*, 1907.

KITTEL und EMELE
Vorlagen für den Unterricht im Fachzeichnen der Uhrmacher.

KLEINLEIN, W. J.
Rules and Practice for adjusting Watches. 12°. 118 pp. *Boston*, 1920.
Practical Balance and Hairspring Work.

KRATZENSTEIN
Annotationes circa constructionem horologii marini. 1750–1.

KREUZER, R.
Zeitgeschichte von Furtwangen. 8°. 277 pp. 1880.
Die Marfel'sche Uhren Sammlung. 4°. 31 pp., 38 pl. 1888.

KRÜGER, P. J.
Theoretisch-praktischer Unterricht in der höheren Uhrmacherkunst, etc. 8°. 126 pp., 31 pl. *Quedlinburg*, 1851.
Neues und vollständiges Handbuch der Uhrmacherkunst. (Translation of Janvier et Magnier.) *Quedlinburg*.

KUCKUCK, J.
Die Uhrenindustrie des Würtembergischen Schwarzwalds. 8°. 175 pp. (B.) *Tübingen*, 1906.

KUEHL, W. H.
Führer durch die gesammte Uhrmacher Literatur. 8°. 30 pp. (B.) (A priced catalogue of books.) *Berlin*, 1892.

LABARTE, Jules
Description . . . de la collection Debruge Dumenil. 860 pp. (V.) *Paris*, 1847.
(Horlogerie et montres, pp. 725–34.)

LABARTE, JULES—*cont.*

Histoire des arts industriels au moyen âge et à l'époque de la Renaissance. 4 vols. and album. (V.) *Paris*, 1864–6.
(Vol. III treats of Horology.)

Handbook of the Arts of the Middle Ages. (*An English translation.*) (V.) *London*, 1855.

LACOMBE, Jacques

Dict. Encycl. des Amusements des Sciences. Vol i., 900 pp. ; Vol. ii, 86 pl. (Includes clocks, etc.) *Paris*, 1792.

LACROIX, Paul

Les arts au moyen âge, etc. (V.)

English translation : The Arts in the Middle Ages and Renaissance. 8°. (G.) *London*, n.d.
(Pp. 138–52 deal with horology.)

LAGRANGE, J. L. de

Réflexions sur l'échappement. 4°.

LANA, F.

Prodromo overo saggio di alcune inventioni nuove premesso all' arte maestra. Fol. 252 pp., 20 pl. (Chapter 9 describes : Oriolo che si muove perpetuamente.) *Brescia*, 1670.

LANCASHIRE WATCH CO.

Its Rise and Progress. 4°. 160 pp. (P.) *Prescot*, 1893.

LANGE, A.

Meisterwerke der Uhrmacherkunst von einst und jetzt. 8 pp., 8 pl. *Glashütte.*

LANGMAN, H. R., and BALL, A. E. J.

Electrical Horology, etc. 8°. 176 pp. (P., S.) *London*, 1923.

LAVAIVRE, J.

Réparations d'horlogerie. 8°. *Paris*, 1905.

LEARNED, W. B.

Watchmaker's and Machinist's Handbook. *Chicago*, 1897.

LEBER, M.

Notice sur l'horloge gothique construite vers 1430 pour Philippe III, dit le Bon, Duc de Bourgogne. 4°. 24 pp., 3 figs. *Vienna*, 1877.

LEBON, E.

Etudes historiques . . . sur l'horlogerie en Franche Comté. 8°. 348 pp. (P.) *Besançon*, 1860.

LECOULTRE, J. E.

Le chronographe et ses perfectionnements. 12 pp., 4 pl. *Lausanne*, 1894.
(Second revised edition.) 8°. 16 pp., 5 pl. (P.) *Geneva*, 1908.

LEDIEU, A.

Les montres marines : description, théorie et perturbations. 8°. 80 pp. 1 pl., 13 figs. 1877.

LEGAL, Fulgence

La marche et la conduite des chronomètres d'après les travaux de MM. Caspari, Lieusson, De Magnac, etc. *Paris*, 1882.

LEIST, G., see Thon, G.

LE NORMAND, L. Séb.
Manuel de l'horloger. 18°. 348 pp. *Paris*, 1830.
Nouvcau manuel complet de l'horloger. *Paris*, 1837.

LE NORMAND et JANVIER
Manuel de l'horloger. (Nouvelle édition par D. Magnier.) 18°. 332 pp., 13 pl. (P., S.) *Paris*, 1850.
Nouvelle édition par L. S. T. 18°. 464 pp., 12 pl. (P.) *Paris*, 1863.
Another edition. 2 vols. 16°. 384 pp. and 268 pp., 12 pl. *Paris*, 1876.
Sixth edition (by E. E. Stahl) : augmentée de l'horlogerie électrique . . . pneumatique, etc. 2 vols. 12°, atlas. (P.) *Paris*.

LE PAUTE, Jean André (1709–89)
Réponse à une lettre du Sieur Caron fils. (It refers to the pin-wheel escapement.) 4°. 12 pp. (P.) *Paris*, 1753.
Traité d'horlogerie, etc. 4°. (G.) *Paris*, 1755.
Second edition. *Paris*, 1760.
Third edition, with supplement in 2 parts. 4°. 375 pp. (P., S.) *Paris*, 1767.

LE ROY, Julien (1686–1759)
Avis contenant les vrais moyens de régler les montres, tant simples qu'à répétition. 8°. 8 pp. *Paris*, 1719.
Second edition. 12°. 8 pp. (P.) 1741.
German translation : Answeisung die einfachen so wohl als Repetir-Uhren wohl zu richten und zu gebrauchen. 8°. 28 pp. (P.) *Dresden*, 1759.
Nachrichten zur Historie der Uhrmacherkunst. (*German translation.*) *Dresden*, 1730.
Mémoires sur différentes parties de l'horlogerie. Being an Appendix to the second edition of Sully's " Règle artificielle du temps." (G.) *Paris*, 1737.

LEROY, L.
Montre ultracompliquée. 8°. 32 pp., 2 pl. *Paris*, [1900].

LE ROY, Pierre (1717–85)
Recherches des longitudes en mer, etc. 4°. 50 pp. *Paris*, 1768.
Exposé succinct des travaux de MM. Harrison et Le Roy dans la recherche des longitudes en mer, etc. 4°. 68 pp. (B., P.) *Paris*, 1768.
Mémoire sur la meilleure manière de mesurer le temps en mer. *Paris*, 1770.
Précis des recherches faites en France depuis l'année 1730 pour la détermination des longitudes, etc. 4°. 52 pp. (P.) *Paris*, 1773.
Suite du précis sur les montres marines. 4°. *Paris*, 1774.
Etrennes chronométriques. Mémoires des Horlogers de Paris. *Paris*, 1760.
(See also M * * *)

LEUTMANN, J. G.
Vollständige Nachricht von den Uhren, etc. 2 vols. 8°. (B., P.) *Halle*, 1717–21.
Another edition. 8°. (P.) *Halle*, 1718–22.
Third edition. 1732.

LEWIS, W. T.
Friction, Lubrication and the Lubricants in Horology. 8°. 96 pp. (P.) *Chicago*, 1896.

LEYBOURN, Wm.
Pleasure with Profit, etc. Fol. (B., G.) *London*, 1694.
(It contains a chapter of 12 pages entitled : Recreations Automatical, giving rules for calculating trains, length of pendulums, equation of time, etc.)

LEYMANN, Otto
Vorschule für Uhrmacher. *Leipsic.*

LIEUSSON, A.
Recherches sur les variations qui affectent la marche des pendules et des chronomètres. 8°. 180 pp. (P.) *Paris*, 1853.

LIMAN, L. de
Nouveau cours d'horlogerie à l'usage des fabricants et des rhabilleurs. 8°. 302 pp., 1 pl. (P.) *Paris*, 1854.
Horlogerie. *Besançon*, 1860.
German translation by T. Bürk : Neues Lehrbuch der Uhrmacherkunst. 4°, 206 pp., 1 pl. 1864.
Second German edition. 1895.
Traité du rhabillage et de la fabrication de l'horloge actuelle. 12°. 436 pp. (P.) *Besançon*, 1861.

LITTROW, L. von
Miscellen über Uhren. 28 pp. 9 figs. 1845.

LOESKE, M.
Die gesammte Literatur über Uhrmacherei und Zeitmesskunde. 16°. (B.)
(A very full bibliography.) *Bautzen*, 1897.
Praktisches Hilfsbuch für Uhrmacher. 8°. 205 pp. *Leipsic*, 1910.
(See also Favarger, Lossier, and Saunier.)

LOSEBY
Improvements in Chronometers. (I.) 1852.
Second edition, by J. J. Farmer. *London*, 1914.

LOSSIER, L.
Etude sur la théorie du réglage des montres, d'après le mémoire de même titre de J. Grossmann. 8°. 135 pp., 3 pl., 41 figs. (P.) *Geneva*, 1890.
Second edition. 8°. 186 pp., 3 pl. (P.) *Geneva*, 1907.
German translation, by M. Loeske : Das Regulieren der Uhren in den Lagen, in Theorie und Praxis. 8°. 230 pp., 3 pl., 48 figs. *Bautzen*, 1892.
Second edition. 8°. 230 pp., 3 pl. (P.) *Bautzen*, 1895.
English translation in the *Horological Journal*, vols. 37–38, 1895–6, by Walker and Barber.

LUDWIG und FRIES
Handbuch der Uhrmacher. 8°. 66 pp., 14 pl.

LUND, J. A.
Synchronised Clocks. 8°. 16 pp. *London*, 1882.

LÜSBERG, Bering
Urmagere og ure i Danmark, et. *Copenhagen*, 1908.

LYON (Ville de)
Mémoire signifié pour la Communauté des maîtres horlogers de la ville de Lyon. *Paris*, 1742.
Museum : Montres et horloges. 1919.

LYON and SCOTT
Evolution of the Timepiece. *Chicago*, 1895.

M.
Der selbstlehrende Uhrmacher. 8°. (G.) *Frankfort*, 1786.

M * * *
Mémoire sur l'horlogerie, etc. (See Rivaz.) 4°. (G.) *Paris*, 1750.
(An anonymous critique by Pierre Le Roy.)

MÄDELER, G.
Handbuch für Landuhrmacher, etc. *Weimar*, 1858.

MAGEE, D. F.
Grandfather Clocks: their Making and Makers in Lancaster County. (Lancaster County Historical Society.) (S.) 1917.

MAGNAC, A. de
Recherches sur l'emploi des chronomètres à la mer. 8°. 90 pp. (P.) *Paris*, 1874.

MAGNIER, see Janvier et Magnier; Normand et Janvier.

MANDEY and MOXON
Mechanick-powers, or the Mystery of Nature and Art unvail'd, . . . together with a Treatise of . . . the making of Clockwork and other Engins. 4°. In 10 books with copper plates. (G.) *London*, 1696.

MANDEY, Venterus
Mechanical Powers . . . with the making of Clockwork and other Engins. 4°. 1709.

MANLEY, Wilhelm
Unterricht von Sackuhren. *Frankfort*, 1728.
Second edition, enlarged. 12°. *Vienna*, 1751.
Third edition. *Nuremberg*, 1761.

MANNI, Domenico M.
Commentarium de Florentinis Inventis. (Chapter 29 is "De horologiorum aliorumque instrumentorum perpetui motus inventionibus." 4°. (B.) *Ferrara*, 1731.

MARFEL Collection, The
(See Kreuzer; Olzinn; Speckhart.)

MARGUERAT, A.
Guide des horlogers, contenant un cours abrégé sur le rhabillage. 8°. 128 pp. (P.) *Geneva*, 1837.

MARSH, E. A.
The Evolution of Automatic Machinery as applied to the Manufacture of Watches. 140 pp., 75 figs. (P.) *Chicago*, 1896.

MARSHALL, P. & Co.
Electric Clocks and Chimes. 8°. 160 pp. (P.) *London*, [1922].

MARTENS, J. H.
Beschreibung der Hemmungen der höheren Uhrmacherkunst. 8°. 272 pp. (Atlas of 24 pl.) *Leipsic*, 1858.
Beschreibung einer neuen freien Chronometer Hemmung, etc. 8°. 4 pl. 1875.
New and enlarged edition. 59 pp., 5 pl. *Freiburg i/B*, 1877.
Second edition. 1890.

MARTIN, B.

Description and Use of a Table Clock upon a New Construction, going by a Weight 8 days. (I.) 1770.

MARTIN, John

Mechanics . . . or the Watch Spiritualised. 8°. (G.) *London*, 1773.

MARTIN, T.

The Circle of the Mechanical Arts. Pp. 563–95, "Watch and Clock-making." *London*, 1813.

For *German translation*, see Poppe.

MARTINELLI, Domenico

Horologi Elementari fatti con l'acqua, con la terra, con l'aria, col fuoco. 4°. 160 pp. (B., P.) *Venice*, 1669.

(Timekeepers actuated by the four elements without wheelwork. See also Ozanam.)

MARTINOT & HAYE

La sphère mobile présentée au Roi. 12°. *Paris*, 1701.

(For a description of this astronomical instrument see Allexandre's "Traité général des horloges," pp. 323–6.)

MARTORELLI, L.

Dissertazione sull' orologio e sull' ore degli antichi Romani, etc. 12°. 34 pp. (P.) *Rome*, 1812.

MASCART, J.

La détermination des longitudes et l'histoire des chronomètres. 8°. 62 pp. (P.) *Besançon*, 1910.

MASKELYNE, Rev. Nevil

Observations on one of Skelton's Clocks. (I.) 1762.

Account of the going of Mr. John Harrison's Watch at the Royal Observatory. (Cf. Harrison.) 4°. 84 pp. (P.) 1767.

An Answer to a Pamphlet, entitled a "Narrative of Facts" by Mr. Thos. Mudge, jr., etc. (Cf. Mudge.) 8°. 172 pp. (P.) *London*, 1792.

MASON, C.

Observations for proving the going of Ellicott's Clock at St. Helena. (I.) 1762.

MASSOTEAU

Instruction pour bien régler les montres. 4°. 4 pp. *Paris*, 1734.

MASSY (of Amsterdam)

His essay on the best means of preserving equality of movement in a clock at sea gained the prize offered by the Académie Royale des Sciences in 1720. 4°. *Paris*, 1722.

(Description in Allexandre's "Traité général des horloges," pp. 359–63.)

MATTHIEU, M.

Rapport sur les instruments d'astronomie . . . et sur l'horlogerie. (Paris Exhibition, 1855.) 4°. *Paris*, 1856.

MAUMENÉ, E. J.

Rapport . . . sur le système d'horlogerie électrique présenté par M. Leroy. 8°. 14 pp., 1 pl. (P.) *Rheims*, 1854.

MAUNDER, E. W.

The Royal Observatory, Greenwich: its History and Work. *London*.

MAYER, C.
A Letter on the going of a New Pendulum Clock made by J. Arnold and set up in the Observatory at Mannheim. (*Translation from the German.*) 4°. 28 pp. (P.) *London*, 1781.

MAYETTE, J.
De la mesure du temps et du réglage des montres. (In *Annales de la Société d'Agriculture.*) (B.) 1890.

MAYHER.
Die astronomische Zeitrechnung der Völker. *Diessen vor München*, 1913.

MAZE-SENCIER, A.
Le livre des collectionneurs. 8°. (V.) 1885.

MEGNIN, G.
Naissance, développement et situation actuelle de l'industrie horlogère à Besançon. 4°. 300 pp. *Besançon*, 1909.

MENADIER, F., see Berthoud.

MENARD, F. D.
(French title not known.) *Italian translation* by G. M. Vendettj: Vero metodo di regolare le mostre, etc. 24°. 20 pp. *Rome*, 1805.

MERLING, A.
Die elektrischen Uhren in allgemein verständlicher Darstellung der Construction, etc. 323 pp., 208 figs. (P.) *Brunswick*, 1884.

MEYER, Jürgen
Die Grundlehren der Uhrmacherkunst. 8°. 1874.
Second revised edition. 8°. 26 figs. *Weimar*, 1874.
Third edition. 12°. 110 pp. (P.) *Weimar*, 1891.

MICHEL, R.
Les premières horloges du Palais pontifical d'Avignon. (In *Ecole française de Rome :* Mélanges, etc. (V.) *Paris*, 1909.

MIGEON, G.
Collection de M. P. Garnier. (In *Les Arts* for March.) (V.) *Paris*, 1906.

MILHAM, W. I.
Time and Timekeepers. 8°. 628 pp. (B., P.) *New York*, 1923.
(Contains an extensive bibliography of books and articles, the latter mostly American.)

MITCHELL, A.
Description of the New Clock of the Cross Steeple of Glasgow. (I.) 1843.

MOINET, Louis (1768–1853)
Nouveau traité général astronomique et civil d'horlogerie théorique et pratique. 8°. 2 vols. 432 and 544 pp. (B., G.) *Paris*, 1848.
Second edition. (B.) [1853.]
Third edition. (P., V.) 1875.
Fourth edition. 8°. 5 vols. *Paris*, 1877.
(Chapter I gives a very lengthy description of De Vick's clock.)

MOLITOR
Anweisung wie Geh-Schlag-Repetir- und Sackuhren richtig berechnet, probirt und tractirt werden. 8°. *Frankfort*, 1762.

MOLLINGER, C.
Kleiner Uhrenkatechismus, etc. 16° *Berlin*, 1818.

MOLYNEUX, W. (F.R.S.)
Sciothericum Telescopium, etc. *Dublin*, 1686.
(Also deals with pendulum watches.)

MOORE (Mrs.), N. H.
The Old Clock Book. (B.) 339 pp., 104 illus. *New York*, 1911.

MORGAN, C. Octavius
Observations on the History and Progress of the Art of Watchmaking. 4°. (G., P.) *London*, 1849–50.
(In *Archæologia*, vol. 33, in two parts.)
List of Members of the Clockmakers' Company, 1631–1732. 4°. (G.) *London* [1883].
In the *Archæological Journal*, vol. 40, no. 138.
Reprinted at Exeter.

MORGAN, J. P.
Catalogue of his Collection of Watches. 4°. (B.) 1912.
(See G. C. Williamson.)

MOUCHEZ (Admiral)
Observations chronométriques faites pendant la campagne de circumnavigation de la "Capricieuse." 8°. 139 pp., 4 pl. *Paris*, 1855.

MOXON, see Mandey.

MUDGE, Thomas (1715–94)
Thoughts on the Means of Improving Watches and more particularly those for use at Sea. 8°. 16 pp. (P.) *London*, 1765.
Register of the Going of Mr. Mudge's First Timekeeper, from April 1780 to May 1781. (G.) [1781.]
Report of the Select Committee of the House of Commons . . . to make Enquiry into the Principles on which Mr. Mudge's Timekeepers have been Constructed. 8°. 160 pp. (P.) *London*, 1793.

MUDGE, Thomas, Junior
A Narrative of Facts relating to Some Timekeepers, constructed by T. M. for the Recovery of the Longitude at Sea, etc. 8°. 106 pp. (G., P.) *London*, 1792.
A Reply to the Answer of the Revd. Dr. Maskelyne, etc. 8°. 172 pp. (G., P.) *London*, 1792.
A Description of the Timekeeper invented by the late Mr. Thomas Mudge. 4°. 328 pp., 9 pl. (P.) *London*, 1799.
Description and Thoughts on the Means of improving Watches. (I.) 1799.

MÜLLER, A. E.
Grundzüge der Kinematik der freien Uhren-Hemmungen, etc. 8°. 48 pp.

MÜLLER, F. C.
Vom Gebrauche der Taschenuhren zu geometrischen Messungen, etc. 8°. 215 pp., 10 pl. *Berlin*, 1777.

MYLES, J. L. W.
The Truth about Watches. *Manchester*, 1905.

NATERMANN, J.
Etude sur l'échappement à ancre. 8°. 35 pp. *Geneva*, 1880.
Second edition.
Third edition. 8°. 60 pp., 3 pl. (P.) *Geneva*, 1909.

NELTROPP (Rev.), H. L.
A Treatise on Watchwork Past and Present. 8°. 322 pp. (G., P., S.) *London*, 1873.
A Catalogue of his Collection of Clocks and Watches presented to the Clockmakers' Company. (G.) *London*, 1895.
Additions to the Collection, etc. 1897.

NEU
Der neue englische Uhrmacher, etc. 8°. 189 pp., 4 pl. (P.) *Frankfort*, 1781.

NICHOLSON, J.
The Operative Mechanic and British Machinist, etc. Second edition. 8°. (Treats of horology.) (G.) *London*, 1825.
German editions. 8°. Plates. *Weimar*, 1826–1834.

NICHOLSON, Wm.
A Journal of Natural Philosophy, etc. 4°. (G.) *London*, 1797–1802.
(Contains articles on horology.)

NILLSON, M. B.
Primitive Time-reckoning, etc. *Lund*, 1920.

NOELDECKEN, W.
Peter Hele, der Erfinder der Taschenuhren. (A Tale for Young People.)

NOTIZKALENDER für Uhrmacher. 1878–

NUTTING, Wallace
The Clock Book. 8°. 304 pp., 250 illus. *Framingham* (*Mass.*), 1924.
(Contains a list of American Clockmakers.)

OLIVIER, T.
Mémoire sur les engrenages. *Paris*, 1839
(In vol. viii of the *Transactions of the Académie des Sciences.*)

OLZINN, A.
Sammlung Marfels. 4°. 2 pp., 17 pl.

ORFF, C. von
Bestimmung der Länge des einfachen Secundenpendels auf der Sternwarte zu Bogenhausen. 4°. 134 pp. 1883.

ORONTIUS, see Finée.

OSTWALD, see Huygens.

OTTO, H.
The First Lever Watch with Free Balance. 22 pp., 1 pl., 24 illus.
(i.e. Mudge's watch made for the King.) *London*, 1931.

OUGHTRED, Wm. (1575–1660)
Opuscula Mathematica. 8°. (Posthumous.) *Oxford*, 1677.
(Calculations of wheelwork—see Clerke.)

OVERALL, W. H. See Atkins.

OVERTON, G. L.
Clocks and Watches. 8°. 138 pp. (P., S.) *London*, 1922.

OWEN, E. W. C.
Use of the Chronometer with suggestion for applying it to correct the Latitude as well as the Longitude. (I.) 1812.

OZANAM, Jacques (1640–1717)
Recréations mathématiques et physiques. (B., G., P.) *Paris*, 1694.
(This contains a French translation of Martinelli's Work (*q.v.*) with observations. Reprinted in Amsterdam in 1696 and 1698, and other editions were published in 1725, 1741 and 1750.)

PAGEL, L.
Marche de la pendule et du chronomètre. 8°. 40 pp. (P.) *Paris*, 1879.
PALMER, Wm.
A New Discovery for finding the Longitude, etc. *York*, 1714.
A Great Improvement in Watchwork . . . for discovering the Longitude at Sea. 8°. 16 pp. (P.) *York*, 1716.
PANCIROLLUS, Guido
Rerum memorabilium antiqua deperdita at nova reperta. (B.) 1599.
(Chapter x treats of clocks. The book was reprinted at Hamburg in 1607 and at Frankfort in 1622 and 1629. An enlarged edition by Heinrich Salmuth in 2 vols. 4° was published at Frankfort in 1660. *Italian translation*, Venice, 1612 ; *French*, Lyons, 1617 ; *English*, 1715 and 1727.)
PARIS (Town)
Musée des Arts Décoratifs : Les nouvelles collections ; 1. Bronzes, pendules, etc. (V.) 1906.
PARKINSON and FRODSHAM
A Brief Account of the Chronometer, etc. 8°. 24 pp. (P.) *London*, 1832.
Change of Rates in Chronometers arising from their transfer from one place to another. (I.) 1833.
Chronometers (Arctic Regions). (I.) 1834.
PARR, Wm.
A Treatise on Pocket Watches. 8°. 56 pp. (G., P.) *London*, 1804.
PARSCH, F. X.
Die Olmützer Kunstuhr. (V.) *Olmütz*, 1900.
PARTINGTON, C. F.
The Clock and Watch Maker's Complete Guide, comprehending the Early History of the Art. 8°. 96 pp., 2 pl. (G., P.) *London*, 1825.
(It contains a bibliography ; in *The Mechanic's Library*.)
PATENT OFFICE (London)
Subject List of Works on Horology. *London*, 1912.
PEARSON, Dr.
Chronometer Escapement (in *Rees' Cyclopædia*).
PEDERSEN, G.
Urmageriet. *Christiania*.
PELLEPORC, Marquis de
Essai sur la partie arithmétique de l'horlogerie. 8°. *Neuchâtel*, 1779.
PENDEL
Collection de mémoires relatives au pendule. 7 pl. *Paris*, 1889.
PENDERELL-BRODHURST, J. G. J.
Article on Clocks in the 11th edition of the *Encyclopædia Britannica*. *Cambridge*, 1911.

PERDUE, E. T.
The Universal Clock Adjuster. 18°. 50 pp. *Richmond, Va.*, 1877.
PERRON, L.
Essai sur l'histoire abrégée de l'horlogerie. 8°. *Besançon*, 1833.
Histoire de l'horlogerie en Franche-Comté. 8°. (G.) *Besançon*, 1860.
PERSEGOL, J. E.
Manuel complet de l'horloger rhabilleur. 12°. 11 pl. 1882.
Nouveau manuel de l'horloger rhabilleur. 18°. 218 pp. (B.) 1895.
Second Edition. 32°. 336 pp. 1907.
PETAU, Père Denis, S.J. (Petavius)
Rationarium Temporum (?)
D. Petavii . . . Opus de doctrina temporum, etc. 2 vols. Fol. (B.) *Paris*, 1627.
(A work on time measurement. It was reprinted in different towns up to 1745. The British Museum possesses ten separate editions beside the above.)
The History of the World ; or an Account of Time . . . continued to 1659. Fol. (B.) *London*, 1659.
PFISTER, A.
Principes de construction des rouages d'horlogerie. Second edition. 8°. 24 pp. (P.) *Geneva*, 1907.
PFLEGHART, A.
Die schweizerische Uhrenindustrie. (B.) 8°. 215 pp. *Leipsic*, 1908.
PHILIPPE, A.
Les montres sans clef, etc. 8°. 308 pp., 3 pl. (G., P.) *Paris*, 1863.
Etudes sur l'horlogerie à l'Exposition de Paris en 1878. 12°. *Geneva*, 1879.
PHILLIPS, M.
Note sur les résultats de la théorie du spiral réglant. 8°. 4 pp., 1 pl. (P.) (?) 1860.
Mémoire sur le spiral réglant des chronomètres et des montres. 4°. 100 pp., 5 pl. (P.) *Paris*, 1861.
Manuel pratique sur le spiral réglant des chronomètres et des montres. 8°. 36 pp. (P.) *Paris*, 1865.
Du spiral réglant conique des chronomètres et de divers autres spiraux. 4°. *Paris*, 1881.
Sundry papers in the *Comptes Rendus de l'Académie.*
PIAGET, H. F.
The Watch, its Construction . . . how to choose it, etc. (P.) *New York*, 1860.
Second edition. *New York*, 1868.
Third edition : The Watch : Handwork versus Machinery. 8°. 37 pp. *New York*, 1877.
PIAZZI, H.
Sull' orologio italiano ed europeo riflessioni. *Palermo*, 1798.
PIERRET, V. A.
Horlogerie, outillage et mécanique. 8°. 79 pp. (P.) 1885.
PIGUET, M.
Histoire de l'horlogerie à la Vallée de Joux. 8°. 87 pp. (B.) *Sentier*, 1895.

Plana, J.
Mouvement du pendule dans un milieu résistant. 4°. 263 pp. 1834.
Planchon, M.
L'Horloge, son histoire rétrospective, pittoresque et artistique. Sm. 4°. 264 pp., 107 illus. (B., P.) *Paris*, [1898].
Second edition, revised by his son-in-law, Paul Brateau. 8°. 232 pp., 150 illus. *Paris*, 1923.
L'Evolution du mécanisme de l'horlogerie depuis son origine. 8°. *Paris*, 1918.
La pendule de Paris. *Paris*, 1921.
Musée rétrospectif: Horlogerie (Exposition Universelle de Paris, 1900). 8°. 152 pp. *Paris*, 1900.
Plank, Stephen
An Introduction to the Only Method for discovering the Longitude. *London*, 1714.
Plantamour, E.
Recherches expérimentales sur le mouvement simultané d'un pendule et de ses supports. 4°. *Geneva*, 1879.
du Plat, E. A.
A Short History of the Various Methods of Measuring Time. 8°. (P.) *London*, 1895.
(A Preface to the Catalogue of the Exhibition of Clocks and Watches at the Royal Aquarium, Westminster, May 15 to June 15, 1895.)
Playtener, H. R.
An Analysis of the Lever Escapement. 88 pp. *Chicago*, 1896.
Pleissner, Paul
Dresden in der Geschichte der Uhrmacherei, etc. 4°. 32 pp., 8 pl. (P.) *Dresden*, 1924.
Ploix, see Delamarche.
Poisson
Mémoire sur les mouvements simultanés d'un pendule et de l'air environnant. *Paris*, 1831.
Pons, H.
Pons' Clock and Watch Manufacture. (I.) 1823.
Timekeepers. (I.) 1846.
Poole, R. L.
Medieval Reckoning of Time. 8°. 48 pp. (S.) *London*, 1918.
Poppe, J. H. M.
Versuch einer Geschichte der Entstehung und Fortschritte der theoretisch-praktischen Uhrmacherkunst. 12°. 90 pp. (G.) *Göttingen*, 1797.
Ausführliche Geschichte der theoretisch-praktischen Uhrmacherkunst seit der ältesten Art den Tag einzutheilen bis an das Ende des 18ten Jahrhunderts. 8°. 564 pp. (B., G., P.) *Leipsic*, 1801.
(It contains an interesting bibliography.)
Praktisches Handbuch für Uhrmacher, Uhrenhändler und für Uhrenbesitzer. 2 vols. 8°. (B.) *Leipsic*, 1799–1800.
New edition. 1810.
Theoretisch-praktisches Wörterbuch der Uhrmacherkunst, etc. 2 vols. 8°. *Leipsic*, 1799–1800.

POPPE, J. H. M.—*cont.*

Der Wecker für Jedermann. *Frankfort*, 1809.

Die Wand- Stand- und Taschenuhren, etc. 1818.

Second edition. 12°. 176 pp. (B.) *Frankfort*, 1822.

Die englische Uhrmacherkunst. (Translation of work by T. Martin.) 8°. 88 pp., 3 pl. *Pesth*, 1819.

(This is Part 1 of "Encyclopädie der mechanischen Künste und Manufacturen," etc.)

Die Uhren und die Uhrmacherkunst auf der höchsten Stufe der jetzigen Vervollkommnung, etc. 528 pp., 12 pl. 1829.

PORTAL, C., and GRAFFIGNY, H. de

Les merveilles de l'horlogerie. 8°. 296 pp., 112 figs. (B., P., S.) *Paris*, 1888.

(The *Journal Suisse d'Horlogerie* says that it is full of technical mistakes. For second edition, see Graffigny.)

POWELL, J.

Depressed State of the Watch Trade. (I.) 1817.

PRAGUE (Town)

Die astronomische Kunstuhr auf dem Altstädter Rathhause in Prag. (By P. F.) 16°. 24 pp., fig. 1891.

PREUD'HOMME, L. B.

Considérations pratiques sur les engrenages des roues et pignons en horlogerie. 4°. *Geneva*, 1778.

Second edition. 12°. (G.) *Geneva*, 1780.

PRINTZ

Försök til en historisk beskrifning om wägg och byrsäcks-uhr. 12°. (P.) *Stockholm*, 1769.

RADDI

Centesimal Clock of E. Cusani compared with Ordinary Clocks. (I.) 1893.

RAILLARD, Claude

Extraits des principiux statuts des maîtres horlogers des années 1544–1719. 4°. *Paris*, 1752.

RAMBAL, J.

Etude théorique et pratique des engrenages pour les pignons de bas nombre. 17 pp., 2 pl., 2 figs. *Geneva*, 1880.

Perfectionnement à la méthode courante de vérification de la grosseur des pignons, etc. 8°. 11 pp. 1882.

Enseignement théorique de l'horlogerie. 8°. 344 pp. (B., P., S.) *Geneva*, 1889.

L'Horlogerie à Genève. (In *Nos anciens et leurs œuvres.*) (V.) *Geneva*, 1903–7.

RAMBAUT, A. A. (Astronomer)

Notes on some Japanese Clocks. 8°. (S., V.) *Dublin*, 1889.

(Reprinted from the *Scientific Proceedings of the Royal Dublin Society.*)

RANDELL, W. L.

Clock Repairing and Adjusting Simply Explained. 8°. 68 pp. *London*, [1923].

Watch Repairing and Adjustment, etc. 46 pp. *London*, 1925.

RECKLESS, R. A.

Verge Watch Cocks. (In *The Connoisseur.*) (V.) *London*, 1914.

RÉDIER, A.
Mémoire sur le pendule conique, etc. *Paris*, 1860.

RÉDIER, J. J.
Notice historique sur l'horlogerie. 18°. (P.) *Paris*, 1845.

REES' Cyclopædia
Article on Horology. (It describes De Vick's clock). (See also Pearson.)

REGNARD, E.
Mémoire . . . sur l'horlogerie électrique. 8°. (P.) *Paris*, 1855.

REID, C. L.
North Country Clockmakers of the 17th, 18th and 19th Centuries. (Reprinted from *Archæologia Aeliana*, 4th Series, vol. i.) 8°. *Newcastle-on-Tyne*, 1925.

REID, Thomas
Horology. (In *Brewster's Edinburgh Cyclopædia.*) (G.) [1819].
Second edition. 4°. 62 pp., 10 pl. (P.) [1830].
Treatise on Clock and Watch Making, Theoretical and Practical. 8°. 476 pp., 19 folding plates. (G.) *Edinburgh*, 1826.
Second edition. *Glasgow*, 1843.
Third edition. (V.) 1847.
Fourth edition. 484 pp. (P.) *Glasgow*, 1849.
Fifth edition. 1852.
Sixth edition.
Seventh edition. 1859.

REIS, F. C.
Prize Essay on Watch-cleaning and Repairing. *Chicago*.

RENDLER, J.
Beschreibung einer astronomischen Uhr . . . von Cajetano eigenhändig verfertiget, etc. 4°. 46 pp., 2 pl. *Vienna*, 1771.

RENOUF, M.
L'Horloge de la Cathédrale de Chartres. (In the *Société française d'Archéologie*, 67th session, p. 334.) 1901.

RESAL, H. S.
Des applications de la mécanique à l'horlogerie. 8°. 240 pp., 4 pl. (P.) *Paris*, 1867–9.

REULEAUX, F.
Erfindung und Herstellung der Uhren. (In the 9th Edition of Buch's " Das Buch der Erfindungen.") (P.)

REVERCHON, L. (Editor of *L'Horloger*)
Translation of Huygens' " Horologium " in the number of November, 1922.

REYMOND, M.
Les pendules du Musée des Arts Décoratifs. (In the *Revue de l'art ancien et moderne.*) *Paris*, 1913.

RICO y SINOBAS, Manuel
He first published in five folio volumes in 1863–7 the " Libros del Saber de Astronomia del Rey D. Alfonso X de Castilla " (1252–84), describing in vol. 4 several methods for measuring time, by sundial, clepsydra, candle, and by wheelwork driven by quicksilver. (B.)

RIDEREAU
Recherches sur le vrai moyen de perfectionner les pendules à secondes, etc. 8°. 72 pp. (P.) *Paris*, 1770.

RIEFLER, S. (See also Finn.)
Präzisions Pendeluhren und Zeitdienstanlagen für Sternwarten, etc. 8°. 72 pp. (P., S.) *Munich*, 1911.

RIES, F. C.
Watch Cleaning and Repairing. *Chicago.*

RIGG, E. (See also Saunier.)
Cantor Lecture on Watchmaking. 39 pp. (In the *Society of Arts Journal.*) *London*, 1881.

RIVAZ, Pierre de
Réponse du Sieur de Rivaz à un mémoire publié contre ses découvertes en horlogerie. (See M. * * *.) 4°. *Paris*, 1751.

ROBERT, Henri
Comparaison des montres marines au barillet dentés avec celles à fusée. *Paris*, 1839.
L'Art de connaître les pendules et les montres, etc. Second edition. 12°. 306 pp., 3 pl. *Paris*, 1849.
Considérations pratiques sur l'huile employée en horlogerie. 8°. 80 pp., 4 pl. *Paris*, 1851.
English translation, by J. Bishop : Practical Considerations on the Oil employed in Clockmaking. 8°. 50 pp. (P.) *Paris*, 1855.
German translation, by J. D. Wayant : Praktische Betrachtungen über das in der Uhrmacherkunst angewandte Oel. *Cologne*, 1855.
Etudes sur diverses questions d'horlogerie. 8°. 274 pp., 4 pl. (G., P.) (This includes the preceding work on lubrication.) *Paris*, 1852.
Recherches sur les perfectionnements à introduire dans les pendules de cheminée. Second edition. *Paris*, 1857.
L'Art de connaître et de régler les montres et les pendules. 16°. 62 pp. (P.) *Paris*, 1867.

ROBERTS, H.
Description of a New Seconds Watch. (I.) 1834.
How to regulate Watches, 1841. Second edition. (I.) 1849.

ROBERTS, N.
Studies on Different Questions in Horology. (I.) 1852.

ROBERTSON, W. B., and WALKER, F.
The Royal Clocks in Windsor Castle, Buckingham Palace, St. James's Palace and Hampton Court. 4°. 42 pp. *London*, 1904.

ROBIN, R.
Mémoire contenant des réflexions sur les propriétés du remontoir. 8°. 62 pp. (P.) *Paris*, 1778.
Mémoire . . . sur les dimensions des horloges de château en gros volume. 4°. 12 pp. 1 pl. (P.) *Paris*, 1784.
Mémoire contenant la description d'un échappement libre ou à détente. Moyen d'employer les montres ordinaires avec le cadran décimal sans changer le mécanisme. 8°. 24 pp., 3 pl. (P.) *Paris*, An II.

ROBINSON, F. S.
The Royal Collections . . . at Buckingham Palace. (V.) (In the *Magazine of Art*.) 1899–1900.

ROBLOT, Ch.
Collection de Cadrans de Montres (Exposition Universelle 1900). 8°. 98 pp. (B., V.) *Paris*, 1900.

ROCKSTROH, H.
Populäre Anweisung wie Thurm- Haus- und Taschenuhren in Absicht auf Gang und Zeit richtig zu stellen sind, etc. 4 pl. 1837.

RODANET, A. H.
L'Horlogerie astronomique et civile, etc. 8°. 208 pp., 26 illus. (B., G., P.) *Paris*, 1888.

ROESSLING AND STOSS
Der Thurmuhren-Bau auf seiner jetzigen Stufe der Vollkommenheit, etc. 8°. 10 pl., 5 tab. 1843.

ROGERS, Inkerman
The Progress of Horology. (A lecture before the British Watch and Clockmakers Guild.) 8°. 24 pp., 1 pl. (P.) *London*, 1921.
Some Ancient Clocks in North Devon. 8°. 8 pp., 3 pl. (P.) *Exeter*, 1923. (In *Somerset and Devon Notes and Queries*.)

ROMBOL
Enseignement théorique de l'horlogerie. *Geneva*, 1889.

RONDOT, Natalis
L'Art et les artistes à Lyon du 14e au 18e siècle. (Posthumous work.) 8°. 352 pp. *Lyons*, 1902.

RORET
Manuel de l'horlogerie. Nouvelle édition augmentée de l'horlogerie électrique. 2 vols. 12 pl.
Manuel de l'horloger rhabilleur.

ROSAT, H.
Les horloges marines.

ROSENFELD, B.
Illustrated Auction Catalogue of Watches, May 2, 1910. (V.)

ROSS, Capt. Sir J.
Observations on Chronometers from Ross's Narrative of his Second Expedition to Arctic Regions, 1829–33. (I.) 1835.

ROSTOCK (Town)
Beschreibung der astronomischen Uhr in der Marienkirche. 8°. 16 pp. (B.) *Rostock*, 1885.

ROUSSEL, M.
Notice sur l'horlogerie. Description d'un échappement nouveau pour les montres. 8°. 27 pp., 1 pl. (P.) *Amiens*, 1841.
Notice historique et technique sur l'horloge du Beffroi d'Amiens. *Amiens*, [1866].

ROUSSIALLE, D.
Etudes sur l'horlogerie à l'Exposition de Paris de 1889. 4°. 67 pp. (B.) *Lyons*, 1890.

ROUYAUX, J. A.
Réflexions sur les chronomètres. *Paris*, 1877.

ROWE, J.
Navigation Improved . . . An Essay on the Discovery of the Longitude by a . . . Horometer. (Letters Patent No. 431 of 1720.) 8°. (P.) *London*, 1720.

RÜFFERT, F. W. (See also Hermann.)
Katechismus der Uhrmacherkunst. Third revised edition. 16°. 236 pp., 229 figs., 7 tab. (B.) (No. 55 of Weber's "Illustrierter Katechismus.") *Leipsic*, 1885.
Fourth edition. *Leipsic*, 1901.
Leitfaden zum Selbststudium für Uhrenbesitzer, etc. 8°. 30 pp. *Berlin.*

S. (F. A.), see F. A. Schmidt.
SALLIER, L'Abbé Claude
Recherches sur les horloges des anciens. (In *Mémoires de Littérature de l'Académie des Inscriptions*, vol. 4.) (P.) 1716.
Also in vol. 5. 1724.
SALMUTH, see Pancirollus.
SALOMONS, Sir David, Bt.
What is the Time? 12°. 40 pp. 1919.
Breguet. 8°. 115 pp., 117 pl. *London*, 1921.
Supplement. 8°. 12 pp., 18 pl. *London*, 1921.
(A monumental work on the author's collection of the watches and clocks of this greatest of watchmakers.)
French edition, enlarged. *Paris*, 1923.
SAMTER, H.
Théorie des Gaussischen Pendels mit Rücksicht auf die Rotation der Erde. *Berlin*, 1886.

SANDOZ, Ch.
Origine et développement de l'industrie horologère à Besançon. 8°. 67 pp. (B.) *Besançon*, 1893.
Les horloges et les maîtres horlogeurs à Besançon. 8°. 88 pp. (B.) *Besançon*, 1905.

SARRANTON, Henri de
L'Heure décimale et la division de la circonférence. *Paris*, 1897.
SAUNIER, Claudius (1816–96)
Traité des échappements et des engrenages. 10 pl. *Paris*, 1855.
Le temps, ses divisions principales : les mesures et leurs usages aux époques anciennes et modernes. 12°. (B.) *Paris*, 1858.
Almanach-Annuaire artistique et historique des horlogers, etc. *Paris*, years 1859–68.
Premier tableau synoptique de l'origine et des principaux progrès de l'art de mesurer le temps, etc. *Paris*, 1865.
Traité d'horlogerie moderne, théorique et pratique. 8°. 840 pp., 21 pl. (P.) *Paris*, 1869.
Second edition. (G.) 1875.
Third enlarged edition. 100 figs., and 30 pl. 1887.
Appendix. 113 pp., 8 pl. 1889.

SAUNIER, Claudius (1816–96)—*cont.*

German translation, by M. Grossmann : Lehrbuch der Uhrmacherei in Theorie and Praxis. 3 vols. *Glashütte*, 1878–9.

Second edition. 4 vols. *Bautzen*, 1891.

Appendix. 1891.

Third edition, by M. Loeske. (P.) *Bautzen*, 1902–5.

Appendix. (P.) [1914.]

English translation, by Tripplin and Rigg : A Treatise on Modern Horology in Theory and Practice. 8°. 860 pp., 21 pl. (G., P.) *London*, 1880.

Second edition. (S.) 4°. 860 pp. 1884.

Guide manuel de l'horloger. 8°. 270 pp. (P.) *Paris*, 1870.

Second edition. (G.) 1873.

Third edition. (B.) 1886

English translation, by Tripplin and Rigg : The Watchmaker's Handbook. (G.) 8°. 512 pp. 1882.

Second edition. *London*, 1888.

Republished 1891.

German translation, by M. Loeske : Praktisches Handbuch für Uhrmacher. 8°. 313 pp. *Bautzen*, 1892.

L'Histoire de la mesure du temps. 1872.

Recueil des procédés pratiques usités en horlogerie. 8°. 1874.

German translation, by M. Loeske : Sammlung praktischer Arbeitsmethoden der Uhrmacherei. 8°. 629 pp. *Bautzen*, 1892.

L'Exposition de 1867 (Horlogerie). (Reprinted from the *Revue Chronométrique.*)

L'Exposition de 1878 (Horlogerie) Rapport du jury. 8°. 30 pp. (P.) *Paris*, 1880.

Les fabriques d'horlogerie américaines. (Article in *La Nature*, p. 4.) *Paris*, 1887.

L'Art de mesurer le temps. (Posthumous.)

SCHADE, A.

Andeutungen über die Fortschritte, welche seit dem Anfang dieses Jahrhunderts bis zum Jahre 1840 in der Uhrmacherkunst gemacht worden sind. 4°. 8 pp. (P.) *Breslau*, [1840].

SCHADE, E. F. F.

Ueber helikoidische Verzahnung und ihre Anwendbarkeit auf den Uhrenbau. 4°. 12 pp., 8 pl. (P.) *Breslau*, 1847.

SCHADE, Ferd.

Vollständiges Uhrmacher Lexicon, etc. 8°. 198 pp., 3 pl. (B., P.) (A bibliography.) *Weimar*, 1855.

Second edition. 1858.

SCHAFHÄUTL

Ueber die Geschichte der Uhren. 4°. 130 pp. 1864.

SCHAIK, W. C. L. van

Ueber die Pendeluhr Galilei's. *Berlin*, 1887.

(In *Zeitschrift für Instrumentenkunde*, pp. 350–428.)

SCHILLING-BAUMANN, J.

Ueber Uhren im Allgemeinen, derer Geschichte und Behandlung. 8°. 19 pp., 12 figs. *Zurich*, 1875.

SCHIREK, C.
Die Uhr in kulturgeschichtlicher und kunstgewerblicher Beziehung. 4° 37 pp., 51 illus. *Brünn*, 1890.

SCHLENKER, M.
Die Schwarzwälder Uhrenindustrie. 8°. 96 pp. *Stuttgart*, 1904.

SCHMIDT, C. H.
Die Englischen Pendeluhren, und zwar Thurmuhren, Hausuhren, etc. 12°. 216 pp., 7 figs. *Weimar*, 1856.
(A translation of Beckett-Denison's Article on Clocks in the 8th edition of the *Encyclopædia Britannica.*)

SCHMIDT, Fr. Aug.
Belehrungen für diejenigen welche Taschenuhren tragen, etc. 12°. 104 pp., 1 pl. (P.) *Liegnitz*, 1795.
Second, revised and enlarged edition. 117 pp. *Leipsic*, 1801.
Beyträge zur Zeitmesskunde für Freunde und Liebhaber von Uhrwerken aller Art. 8°. 8 pl. (P.) *Leipsic*, 1797.

SCHMIDT, G.
Ueber astronomische Uhren. *Berlin*, 1868.

SCHMIDT, J. J.
Nachricht und Vorschläge von Verbesserung der Gehe- und besonders Schlage-Uhren. 8°. *Brandenburg*, 1795.

SCHMIDT, O.
Uhrmacherhandwerk dargestellt, etc. (B.) *Leipsic*, 1896.
(In *Verein für Socialpolitik.*)

SCHMIDT-WEISSENFELS
Zwölf Uhrmacher : biographisch novellistische Bilder. 12°. 154 pp.

SCHNEEBELI, Dr. H.
Die elektrischen Uhren, etc. 8°. *Zurich*, 1878.
(In *Technische Mittheilungen*, No. X.)

SCHOOF, G. W.
Lever Escapement for Watches and Marine Chronometers. (I.) 1874.
Improvements in Clocks and Marine Chronometers. 8°. 16 pp. (P., S.) *London*, 1898.

SCHOTT, Gaspar (S.J.)
Technica curiosa. 4°. (B.) *Würzburg*, 1664.
(Book 6 describes a remarkable planetary clock. Book 9, entitled *Mirabilia chronometrica*, treats of wheels, pendulums, springs, etc. A précis of these books is given in Père Allexandre's work (*q.v.*), pp. 300–9.)
Second edition. 4°, in two parts. (G.) *Würzburg*, 1687.

SCHOTT, K.
Die Schwarzwälder Uhrmacherei. 8°. 48 pp. 1873.

SCHOUFFELBERGER, H.
Dix tableaux pour déterminer la grandeur des roues et pignons d'horlogerie. 4°. 8 pp. 2 pl. (P.) *Chaux de Fonds*, 1879.
English translation, by T. Gribi : Wheels and Pinions : How to determine their exact size. *Chicago*, 1896.

22

SCHREIBER, Emmanuel

Geschichte der Uhrmacherkunst von den ältesten Zeiten bis auf unsere Tage. *Weimar*, 1850.

(The Work of Barfuss re-edited.)

Vollständiges Handbuch der Uhrmacherkunst, etc. Third edition. 647 pp. *Weimar*, 1865.

Fourth edition. 1869.

Also new edition by H. Grosch (*q.v.*)

SCHUCKBURG, E.

Anweisung zur Behandlung und Regulierung aller Thurm-Wand-Stutz- und Taschen-Uhren. 12°. *Nordhausen*, 1847.

SCHULTE

Lexikon der Uhrmacherkunst. Second edition. 8°. 984 pp. *Bautzen*, 1902–3.

SCHULTZ, W.

Der Uhrmacher am Werktisch. 8°. 344 pp., 5 pl. (P.) *Berlin*, 1902.

Second edition, enlarged. *Berlin*, 1908.

French translation, by Ch. Gros : L'Horloger à l'établi. 8°. Illus. *Paris*, 1904.

Second edition. Third edition after the second German edition. 8°. 428 pp. *Paris*, 1908.

Unsere Zeitmesser und ihre Behandlung, etc. 8°. 32 pp. (P.) *Berlin*, 1900.

English translation, by Hertz : Our Watches and Clocks and how to manage them. 8°. 32 pp. *London*, 1901.

SCHULZE

Sur les horloges à pendule. 4°.

SCHUMANN, R.

Ueber den Gang der Pendeluhr F. Dencker. *Leipsic*, 1888.

SCHWANATUS, W.

Der Taschenuhr-Gehäusemacher. 8°. 1884.

Second edition. 88 pp. *Bautzen*, 1894.

Repairing Watch Cases. 40 pp. *Chicago*.

SCHWEDELER-MEYER, E.

Die Uhren im Legat des Freiherrn Heinrich von Liebig (im Nordböhmischen Gewerbemuseum). (V.) *Reichenberg*, 1906.

SCHWILGUÉ, C.

Anleitung zur Behandlung der Thurmuhren, etc. 8°. *Strasburg*, 1846.

Description abrégée de l'horloge astronomique de la Cathédrale de Strasbourg. *Strasburg*, 1852.

Notice sur la vie, les travaux, etc., de . . . J. B. Schwilgué. 8°. 286 pp., 1 pl. (P.) *Strasburg*, 1857.

Second edition. 8°. 188 pp., 1 pl. (P.) *Strasburg*.

SCORESBY, W., Junr.

Errors in the Sea Rates of Chronometers arising from the Magnetism of their Balances. (I.) 1882.

SCOTCHFORD, T. C.

A Treatise on the Detached Lever Escapement and a Dissertation on Watch Motions. 8°. 64 pp., 12 pl. (P.) *London*, 1866.

SCOTT, see Lyon

SEEBASS, L. A.
Praktische Anweisung zur Behandlung und Reparatur der Thurmuhren. 5 pl. *Quedlinburg*, 1846.
Gründlicher Unterricht alle an den Thurmuhren vorkommenden Fehler . . . abzuhelfen. 5 pl. *Quedlinburg.*

SÉGUIER, M.
Echappement à ancre. (I.) 1845.
Rapport sur les montres marines ou chronomètres de M. Robert. 4°. 12 pp., 1 pl. *Paris*, 1846.

SERVIÈRE, N. Grollier de (1593–1686)
Recueil d'ouvrages curieux . . . ou description du Cabinet de M. de Servière. 4°. 120 pp., 83 pl. (Published by his grandson.) *Lyons*, 1719.
Second edition. *Lyons*, 1733.
Revised and enlarged edition. 4°. 157 pp., 88 copper plates. (G.) *Paris*, 1751.
(A remarkable work illustrating a number of clocks of Servière's invention, mostly on the principle of a ball descending an inclined plane; also the floating tortoise.)

SHADWELL, C. F. A.
Notes on the Management of Chronometers, etc. 1855.
Second edition. 8°. 244 pp. (P.) *London*, 1861.

SHARP, Isaac
Daniel Quare. (V.) (In the *Friends' Quarterly Examiner.*) January 1900.

SHEPHERD, C.
On the Application of Electro-magnetism as a Motor for Clocks. 8°. (P.) *London*, 1851.

SHERWOOD, N. B.
Watch Repairing. *Chicago.*
Watch and Chronometer Jeweling. *Chicago.*

SHORT, James (F.R.S.)
An Account of the Proceedings in order to the Discovery of the Longitude, etc. (Published anonymously.) *London*, 1763.
(Deals with Harrison's attempts.)

SIEVERT, H.
Leitfaden für Uhrmacher-Lehrlinge. 8°. 263 pp., 64 figs. *Berlin*, 1879.
Many editions—the Seventh. 360 pp. (P.) 1902; eighth *Berlin*, 1906.
Dutch translation: Second edition: Leerboek voor den horologemaker. *Bolsward*, 1905.
French translation. 339 pp. *Paris*, 1913.

SIVAN, C.
L'Exposition historique d'horloges à Nuremberg en 1905 . . . Notice concernant . . . Pierre Henlein. 8°. 72 pp., 5 pl. (P.) *Geneva*, 1906.

SMITH, John
Horological Dialogues, showing the nature, use and right managing of clocks, etc. 12°. 134 pp. (B., G., P.) *London*, 1675.
(The first English work on horology, under initials J.S.)

SMITH, John—*cont.*

Of the Unequality of Natural Time with its reason and causes; together with a table of the true Æquation of Natural Dayes . . . drawn up for the adjusting . . . of Pendulum Clocks. 8°. 42 pp. (B., P.) *London*, 1686.

Horological Disquisitions concerning the Nature of Time, etc. 12°. 92 pp. (B., G.) *London*, 1694.

(The copy in the Guildhall Library has MS. notes by Cumming and Wallace and a biography of Smith.)

Second edition, "much corrected." 8°. 104 pp. (B., P.) *London*, 1708.

SMITH, John

Old Scottish Clockmakers. (P.) *Edinburgh*, 1903.

Revised and enlarged edition. 8°. 451 pp., 25 pl. (P., V.) *Edinburgh*, 1921.

A Handbook and Directory of Old Scottish Clockmakers from 1540 to 1850. 8°. 16 pp., 1 pl. (P.) *Edinburgh*, [1903].

SMITSON, G.

Hülfsbüchlein für Uhrenbesitzer, oder die Kunst seine Uhren selbst zu repariren, etc. 8°. Figs. *Quedlinburg*, 1825.

Anweisung für Uhrenmacher und sonstige Liebhaber genauer Uhren.

SMYTH, Admiral W. H.

Description of an Astrological Clock made by Jacob Zech of Prague in 1525, prefaced by notes on Old Clockwork. (In *Archæologia*, vol. 33, 1849.) Reprint 4°. 30 pp., 1 pl. (P.) *London*, 1848.

Description of the Dover Castle Clock on its First Discovery. (In *Archæologia*, vol. 34, 1852.)

SOCIÉTÉ POUR L'AVANCEMENT DES ARTS

Description des échappements les plus usités en horlogerie. 4°. 44 pp., 12 pl. *Geneva*, 1831.

Second edition. (P.) *Geneva*, 1851.

SOIL, Eugène

Exposition d'art ancien. (Includes clocks.) (V.) *Tournai*, 1899.

SŌTOME, Kiyofusa

Studies on Astronomical Timekeepers and Time-preserving Systems. *Tokyo*, 1921.

(In vol. 5 of the *Annals of the Observatory of Tokyo.*)

SPANJE, T. v.

Handboek voor den horologemaker. *Leyden*, 1864.

SPECKHART, G.

War Peter Hele der Erfinder der Taschenuhren. 1894.

Kunstvolle Taschenuhren der Sammlung Marfels. 8°. (B., S.) *Berlin*, 1904.

SPITZER (F.) Collection. Sale Catalogue, see Bonnaffé.

SPITZER, S.

Die Uhr. Ein Beitrag zur Kulturgeschichte der Alten. 1885.

SPRENGEL, P. N.

(In "*Handwerke und Künste in Tabellen.*") *Berlin*, 1778–95.

Vol. 7. Kunst des Grossuhrmachers.

Vol. 8. Kunst des Kleinuhrmachers.

(Ferd. Schade says they are the best treatises that had been written on the subject.)

STAHL, E. E. (see Le Normand et Janvier)
Nouveau manuel complet de l'horloger. 2 vols. 8°. 1896.
STECHERT, C.
Das Marine-Chronometer, etc. 1894.
STELLE, J. P.
American Watchmaker and Jeweler. *New York*, 1868.
Second revised edition. 12°. 98 pp. *New York*, 1873.
STEYRER, F.
Geschichte der Schwarzwälder Uhrmacherkunst, etc. 8°. (B.) *Freiburg*, 1796.
STÖCKEL, H. F. A.
Anweisung für Liebhaber des Uhrenbaues. 8°. *Nuremberg*, 1802.
STÖCKEL, J. M.
Praktische Lehre . . . über die Uhrenbau in seinem ganzen Umfange, etc. 8°. 214 pp., 16 pl. (P.) *Munich*, 1820.
STONE, E., see Bion, N.
STOSS, see Roessling.
STRANDT, A.
Beschreibung der beruehmten Uhr und Kunstwercke am Altstädter Rathhaus, und auf der kgl. Sternwarte zu Prag. 4°. 6 pl. *Dresden*, 1791.
STRASSBURG (Town)
Beschreibung des Strassburger Münsters und seiner Glocke. 1617.
STRASSER, L., see M. Grossmann.
STRÖBEL, August
Alte Uhren. (In *Internationale Sammler-Zeitung*, April 1914.) *Vienna*.
SULLY, Henry (1680–1728)
Abrégé de quelques règles pour faire un bon usage des montres, etc. (B.) *Leiden*, 1711.
Description d'une montre de nouvelle construction présentée à l'Académie Royale des Sciences. *Paris*, 1716.
German translation. *Lemgo*, 1716.
Regle artificielle du Temps, ou Traité de la division naturelle & artificielle du Temps : Des Horloges & des Montres de differentes Constructions : De la maniere de les connoître & de les regler. 8°. 138 pp., 2 pl. (B., P.) *Paris*, 1717.
Second enlarged edition, by Julien Le Roy. 12°. 262 pp., 3 pl. (G., P.) *Paris*, 1737.
German translation, by A. Charles : Unterricht von der Eintheilung der Zeit, etc.
Second edition. 8°. 264 pp., 8 pl. (P.) *Lemgo*, 1754.
Description abrégée d'une Horloge d'une Nouvelle Invention pour la Juste Mesure du Tems sur Mer . . . et une dissertation sur la nature des tentatives pour la Découverte des Longitudes dans la Navigation, & sur l'usage des Horloges, pour la Mesure du Tems en Mer. 4°. 48 pp., 3 pl. (B., G., I., P.) *Paris*, 1726.
(On p. 44 Sully says that the English edition is ready for the press.)
Méthode pour régler les montres et les pendules. 12°. (B.) *Paris*, 1717.
Second edition. (I.) *Paris*, 1728.

SULLY, Henry (1680–1728)—*cont.*
German translation : Nöthige Unterricht und Regeln für diejenigen welche Sackuhren tragen. *Augsburg*, 1754.
New edition. *Augsburg*, 1804.
Kurtzer Unterricht wie man sich der Sackuhren wol gebrauchen solle, etc. 12°. 58 pp. (P.) *Dettingen.*

SUTTER, Capt.
La Montre ancienne à coq. 35 pp., 3 pl. *Grenoble*, 1911.

SWAN, C.
Le travail des métaux employés par l'horloger et mécanicien. 2 vols. 8°. (P.) *Geneva*, 1893–5.
L'Exposition historique d'horlogerie à Nuremberg en 1905. . . . Précédé d'une notice concernant Pierre Henlein. 8°. 72 pp., 5 pl. (P.) *Geneva*, 1906.

SWINDEN, J. H. van
Verhandeling over Huygens als uitfinder der Slingeruurwerken. 1817. (In the *Transactions of the Kon. Nederl. Institut*, Part 3.)
English abridged translation : History of the Invention of Pendulum Clocks by Christian Huygens. 1822.
(In the *Edinburgh Philosophical Journal*, vol. 6, pp. 197–213, and vol. 7, pp. 35–47.)

TAKABAYASHI, Hyōé
Tokei Hattatsu-shi. (Development of Clocks.) 1927.

TAVAN
Description des échappemens les plus usités en horlogerie. 4°. 12 pl. *Geneva*, 1851.

TAYLOR, Rev. A.
The Watch and the Clock. 8°. 16 pp. *New York*, 1883.

TERRY, H.
American Clockmaking, its early history and present extent. 8°. 26 pp. *Waterbury*, 1870.

TESTORF, F.
Die Elektricität als Antriebskraft für Zeitmesserinstrumente. 8°. 216 pp. (P.) *Halle*, 1910.

THACKER, Jeremy
The Longitudes Examin'd, etc. *London*, 1714.

THIOUT, Antoine (1692–1767).
Traité de l'horlogerie mécanique et pratique, etc. 2 vols. 4°. 430 pp., 91 pl. (B., G., P., S.) *Paris*, 1741.

THOMSON, Adam
Time and Timekeepers. 12°. 207 pp. (G., P., V.) *London*, 1842.

THON, G.
Der Uhrenfreund, oder fassliche Anleitung alle Arten mechanischer Uhren . . zu prüfen, etc. Second edition. Third edition by G. Leist. 1869.

THURY, M.
Etudes sur les conditions actuelles de la fabrication de l'horlogerie à Genève. 8°. *Geneva*, 1874.
Un nouveau pendule compensateur et l'alliage de Sèvres. 8°. 15 pp. (B.) *Geneva*, 1897.
" TIC-TAC "
Les coqs de montres et leur classement historique. 16 cuts and 50 illus. (V.) *Paris*, 1900.
TIRABOSCHI, Girolamo
Storia della Letteratura Italiana. 11 vols. 4°. (B.) *Modena*, 1772–95.
(Contains an account of Dondi's Padua Clock. There are many editions and translations. See article by Adml. W. H. Smyth in *Archæologia*, vol. 34, p. 13.)
TITZER, K.
Hilfsbuch für Uhrmacher. 8°. 25 pp.
TOBLER, A.
Die elektrischen Uhren, etc. 8°. 208 pp. (P.) *Vienna*, 1883.
(In Hartleben's *Electrotechn. Bibliothek.*)
Second edition, by Zacharias: 8°. 272 pp., 2 pl. (P.) *Vienna*, 1909.
L'Horlogerie électrique. 8°. 152 pp. (B.) *Paris*, 1891.
TRAUTMANN, F.
Kunst und Kunstgewerbe. (B.) *Nördlingen*, 1869.
(Contains a list of clockmakers, pp. 380 ff.—inaccurate.)
TRIPPLIN, Julien. (See also Saunier)
Watch and Clockmaking in 1889 . . . with a view of the British Section at the French International Exhibition. 8°. 142 pp. (G., P., S.) *London*, 1890.
Rapport sur l'état de l'horlogerie anglaise en 1900. 8°. 7 pp. *London*, 1900.
Report on the Clocks, Watches and Timepieces at the Paris Exhibition of 1900. 8°. 6 pp. (B.) *London*, 1901.

" UHRMACHER " (Anonymous)
Der neue Englische Uhrmacher, etc. 8°. *Frankfort*, 1781.
Der selbstlehrende Uhrmacher, etc. *Frankfort*, 1786.
Second edition. 8°. *Frankfort*, 1791.
Der belehrende Uhrmacher, oder kurze Belehrung für die welche Taschen- Wand- Tisch- und Schlaguhren besitzen. 8°. *Leipsic*, 1800.
ULRICH, J. G.
Ulrich's Chronometers, etc. (I.) 1846.
UNGERER, T.
L'Horloge astronomique de Strasbourg. 1922.
Les horloges d'édifice, leur construction, leur montage, leur entretien. 8°. 340 pp. *Paris*, [1926].

VAN KEUREN, W. J.
Repairing American Watches. 48 pp. *Chicago*.

VANSSAY, P. de, see Fichot.

VELTMANN, H.

Handschriftliche Aufzeichnungen über einige alte jetzt verschwundene Uhrwerke der Stadt Osnabrück. 72 pp. 1890.

Jost Bodeker der Erfinder des Pendels, nicht Galilei. 40 pp. (Contains Bodeker's description of his circular pendulum.) *Wetzlar*, 1917.

Das Meisterstück des Strassburger Uhrmachers Isaak Habrecht (1578). 4 pp., illus.

Les Habrecht. 55 pp., illus. *Strasburg*, 1925.

VENDETTJ, G. M., see Menard.

VENICE (Town)

Relazione storico-critica della Torre dell' Orologio di San Marco. 8°. 7 pl. 1860.

VÉRITÉ, A. L.

Notice descriptive de l'horloge astronomique de l'église cathédrale de Besançon. 8°. 32 pp., 1 pl. (P.) *Besançon*, 1876.

Description de l'horloge monumentale de la Cathédrale de Beauvais. *Amiens*, 1868.

Second edition. 1876.

VIAL, E., et CÔTE, Ch.

Les horlogers Lyonnais de 1550 à 1650. 8°. 253 pp., 12 pl. *Mâcon*, 1927.

VIGNIAUX

Horlogerie pratique à l'usage des apprentis et des amateurs. 8°. (B., G.) *Toulouse*, 1788.

Second edition. 1802.

Traité élémentaire d'horlogerie. 12°. *Toulouse*, 1800.

Second edition enlarged. 8°. 446 pp., 14 pl. (P.) *Toulouse*, 1802.

VILLARÇEAU, A. J.

Notice sur les travaux de Yvon de Villarçeau. 4°. (B.) *Paris*, 1855.

VILLARÇEAU, Yvon de

Recherches sur le mouvement et la compensation des chronomètres. 4°. 1 pl. 1862.

Exposé concernant les régulateurs isochrones à ailettes. 1 pl. *Paris*, 1883.

VINCE, Rev. S.

Fundamental Properties of the Lever. (I.) 1793.

VOGEL, C. F.

Praktischer Unterricht von Taschenuhren, etc. 8°. 460 pp., 6 pl. (P.) *Leipsic*, 1774.

Versuche, Vortheile, Grundsätze und Regeln zur Erreichung der möglichsten Vollkommenheit der Taschenuhren, etc. 12°. 250 pp., 1 pl. (P.) *Meissen*, 1790.

(Adapted from Berthoud's " Essai sur l'horlogerie," with additions and comments.)

VOLLGRAFF, Dr. J. A.

Christian Huygens : L'Horloge à pendule de 1656 à 1666. *Leiden*, 1931. (Extrait des " Œuvres complètes de Chr. Huygens," vol. xvii, in preparation.)

VULLIAMY, B. L.
Improved Method of Constructing and Theory of Dead-beat Escapements for Clocks. (I.) 1822–3.
Some Considerations on the Subject of Public Clocks, etc. 4°. 16 pp. (G., P., V.) *London*, 1828–30.
Supplement to the above. 8°. 32 pp. (P.) *London*, 1830.
Second edition. 8°. 2 parts. (P.) *London*, 1831–7.
Supplement to the above. 8°. 52 pp. (P.) *London*, 1844.
Advantage attending New Mode of Construction of a Turret Clock. (I.) 1831.
Testimonials Respecting the Clocks made by B. L. Vulliamy. 1837.
On the Construction and Theory of the Dead-beat Escapement for Clocks. 8°. 36 pp., 5 pl. (P., V.) *London*, 1846.

WADE, G. A.
The Picture Backs of Old Watches. (V.) *London*, 1900.
(In the *English Illustrated Magazine*.)

WAGNER, J.
Nouvelles théories et nouveaux principes géométriques de divers échappements . . . adoptés en horlogerie. (Vol. xix. of *Comptes rendus de l'Académie des Sciences*.) *Paris*, 1844.
Anweisung zur Konstruktion neuer verbesserter Thurm Uhren, etc. (From a French work.) 2 pl. *Quedlinburg*, 1846.

WAGNER, J. (neveu)
Mémoire sur le pendule et le balancier, etc. 8°. (S.) *Paris*, 1867.

WÅHLIN, Theodor
Horologium mirabile lundense ; det astronomiske uret i Lunds Domkyrka. 4°. 217 pp. *Lund*, 1923.

WALKER and BARBER, see Lossier.

WALKER, F., see Robertson, W. B.

WALTENHOFEN, A. v.
Ueber die elektrische Uhr von G. Riebeck. 4°. 6 pp., 1 pl. *Prague*, 1879.

WARTMANN, E. F.
Notice historique sur les inventions . . . faites à Genève. (Includes watchmaking.) 8°. *Geneva*, 1873.

WATCHMAKER'S LIBRARY
A Collection of the best Articles from Trade Journals. 290 pp. *Chicago*.

WATCHMAKER'S and JEWELER'S
Practical Receipt Book. 132 pp., illus. *Chicago*.

WAUCHOPE, R.
Plans for ascertaining Rates of Chronometers by an Instantaneous Signal. (I.) 1836.

WAYANT, J. D. (see Robert, Henri).

WEBSTER, M. R. (see Cescinsky).

WEHRLE, E.
Pendule ancienne et curieuse. (In *Annales de la Société d'Archéologie*, vol. xvi, p. 205.) (V.) *Brussels*, 1902.

WEINITZ, F.
Die Landschaftsuhr . . . im Schloss zu Altenburg. (V.) 1910

WELPER, E.
Neu-vermehrte Welperische Gnomonica. *Nuremberg*, 1708.
(Contains a translation of Derham's "Artificial Clockmaker" as appendix.)

WENDEROTH, G. F. W.
Kurze und fassliche Anweisung zu einer zweckmässigen . . . Behandlung der Taschenuhren. 8°. 1 pl. *Eisenbach*, 1804.
(Instructs how to set the watch *by the sun*, and gives equation tables.)

WETHERFIELD, D. A. F. (Collection : see Britten, F. J.)
The Wetherfield Collection of 222 Clocks (to be sold by W. E. Hurcomb, on May 1, 1928, all illustrated). 4°. 95 pp. *London*, 1928.

WHEELER, M.
A Letter . . . concerning a Clock Descendant on an Inclined Plane. 8°. (P.) *Oxford*, 1684.

WHITE, M.
On the Erratic Propensities of Chronometers. *Jersey*, 1830.

WIEDEMANN, E., und HANSEN, F.
Uhr des Archimedes und zwei andere Vorrichtungen. 4°. 40 pp. (B.) 1918.

WIJK, F. J. A. van
Oude Klokken. 4°. 24 pp. *Amsterdam*, 1913.

WILKINSON, T. J.
The Lever Escapement : Test and Errors ; their Detection and Correction. *Philadelphia*, 1916.

WILLARD, J. W.
A History of Simon Willard, Inventor and Clockmaker. Fol. 142 pp. 38 pl. (B., P.) *Boston*, 1911.

WILLIAMSON, Dr. G. C.
Catalogue of the Pierpont Morgan Collection of Watches. (Privately printed. Issued in several sumptuous styles.) (V.) 1912.
Behind my Library Door. (Chaps. xi and xii.) (V.) *London*, 1921.

WILLIS, Prof. R.
Principles of Mechanism. (B.) 1841.
Second edition. 1870.

WINS, Alfonse
L'Horloge à travers les âges. 8°. 373 pp. (P.) *Mons*, 1924.
(Contains a bibliography.)

WOLBRECHT, G.
Die Uhrmacherkunst. (See Anonymous Works.) *Leipsic*, 1828–9.

WOOD, E. J.
Curiosities of Clocks and Watches. 8°. 443 pp. (G., P., S., V.) *London*, 1866.

WOODLEY, W.
Woodley's Patent Timepiece. (I.) 1833.

WORKSHOP NOTES
For Jewelers and Watchmakers. *New York*, 1892.

Wright, T. D.
Technical Horology. 8°. 114 pp.
Wyatt, Sir M. D.
The History of the Manufacture of Clocks. 12°. 32 pp. (P., V.) *London*, 1868.

Zacharias, J. (See also Tobler).
Electrotechnik für Uhrmacher. 318 pp., 4 pl. (P.) *Berlin*, 1908.
Zarbl, see Dietzschold.
Zerella e Icoaga, M.
Tratado general y matematico de reloxeria, etc. *Madrid*, 1789.
(Clockmaker to Ferdinand VI. The book is inaccurate.)
Zöllner, F.
Zur Geschichte des Horizontalpendels. *Leipsic*, 1872.
Zwerger, M.
Der Schwingungsmittelpunkt zusammengesetzer Pendel; historisch-kritische Untersuchung. 129 pp. 1889.
Zwink, M.
Die Pendeluhren im luftdicht verschlossenen Raume, etc. 32 pp. 1888.

Some Periodicals dealing with Clockwork

Allgemeines Journal der Uhrmacherkunst. *Leipsic*, 1876– .
Allgemeine Uhrmacher Zeitung. *Leipsic*, 1888– .
American Horological Journal. *New York*, 1867–73.
American Jeweler.
Annuaire des Fabricants d'Horlogerie de France. *Paris*, 1899– .
Annuaire du Bureau des Longitudes. *Paris*, 1796– .
Annuaire Suisse de l'horlogerie, etc. *Geneva*.
Archæological Institute, Journal of
Argus, L' : Livre d'or de l'horlogerie (Annual). *Besançon*.
Christiaan Huygens Vakblad voor Uurwerkmakers. 1891– .
Chronometer, The *Toledo, U.S.A.*, 1883– .
Chronometro, El *Barcelona*.
Clock and Watchmaker.
Deutsche Uhrmacher Zeitung. *Berlin*, 1877– .
Edinburgh Philosophical Journal.
English Mechanic. 1865- .
English Watchmaker, Jeweller and Silversmith. 1881– .
Fabricant français, etc., Le *Besançon*, 1919– .
Fédération horlogère Suisse. *Chaux-de-Fonds*, 1886– .
France horlogère, La *Besançon*, 1901– .
Gill's Technical Repository.
Giornale degl' Orologiaij.
Glasgow Mechanic's Magazine.
Hora.

SOME PERIODICALS DEALING WITH CLOCKWORK—*cont.*

Horloger, L' *Paris*, 1905– .
Horologemaker, De 1884– .
Horological Journal, The *London*, 1858– .
Internationale Rundschau für die gesammte Uhrenindustrie. 1892–3 only.
Jeweler's Circular and Horological Review. *New York*, 1870– .
Journal of Science and the Arts (Quarterly).
Journal Suisse d'Horlogerie. *Geneva*, 1876– .
Keystone, The *Philadelphia*
Kosmos. *Nuremberg.*
Leipziger Uhrmacher Zeitung. 1906– .
London Mechanics' Register.
Mechanics' Magazine.
Moniteur de l'horlogerie et de l'électricité. 1889– .
Nautical Magazine.
New London Mechanic's Register.
Nicholson's Operative Mechanic.
Oesterreichisch-Ungarische Uhrmacher Zeitung. 1881– .
Popular Mechanics. *Chicago.*
Practical Watch and Clockmaker. *London*, 1928– .
Register of the Arts and Sciences.
Repertory of Arts.
Revue Chronométrique, La *Paris*, 1855– .
Revue des Bijoutiers, Orfèvres et Horlogers.
Revue horlogerie de Besançon. 1891– .
Revue internationale de l'horlogerie. *Chaux-de-Fonds*, 1900– .
Rundschau für Deutschland's Uhrmacher, Goldschmiede und Juweliere. 1891– .
Schweizer Uhrmacher Journal. 1888– .
Schweizerische Uhrmacher Zeitung. 1879– .
Smith's Mechanic.
Society of Arts Transactions.
Süddeutsche Uhrmacher Zeitung. *Augsburg*, 1889– .
Svensk Urmakeri-Tidning. *Stockholm*, 1887– .
Tidsskrift for Uhrmagere. *Copenhagen*, 1881– .
Time and its Keepers. (1 vol. only.) (P.) *London*, 1878.
Tribune chronométrique, La etc. (Dubois). *Paris*, 1852.
Uhrmacherkunst. 1866 only.
Watchmaker, Jeweller and Silversmith, The *London*, 1875– .
Watchmaker and Metalworker. *U.S.A.*
Zeitschrift für Gross- und Kleinuhrmacher. *Weimar*, 1845–8 *and* 1853–5.
Zeitung für Uhrmacher (Schmultz). *Leipsic*, 1841–3.

INDEX

C

D

I

P

Q

R

S